Innovations in Agriculture

Diffusion to Disruption to Degrowth

Dr. R. M. Prasad retired as Associate Director of Extension from Kerala Agricultural University (KAU). He had his Bachelors and Masters' degree in Agriculture from KAU and Ph.D.in Agricultural Extension from UAS, Bangalore. He served KAU for more than 32 years in various positions and was the Professor and Head of Communication Centre and also Head of Central Training Institute, KAU. His areas of interest are Technology transfer, S & T interventions, Convergence, HRD and Skill development. Dr. Prasad is the Life Member of Indian Society of Extension Education (ISEE) and Indian Society of Training and Development (ISTD).

Dr Prasad had served as Advisor, Government of Meghalaya, Agricultural Consultant for ADB and as Senior Fellow in NIRD, Hyderabad. He also served as Training Specialist in two European Union funded projects in Kerala. He was also the National Facilitator of MANAGE, Hyderabad and had served as Member of Sub Group on Agricultural Extension of the Planning Commission, GOI for the 11th and 12th Five Year Plans.

Innovations in Agriculture

Diffusion to Disruption to Degrowth

– Author –

R. M. Prasad

2023

Daya Publishing House®

A Division of

Astral International Pvt. Ltd.

New Delhi – 110002

ISBN: 978-93-5461-675-4 (HB)

Publisher's Note:

Every possible effort has been made to ensure that the information contained in this book is accurate at the time of going to press, and the publisher and author cannot accept responsibility for any errors or omissions, however caused. No responsibility for loss or damage occasioned to any person acting, or refraining from action, as a result of the material in this publication can be accepted by the editor, the publisher or the author. The Publisher is not associated with any product or vendor mentioned in the book. The contents of this work are intended to further general scientific research, understanding and discussion only. Readers should consult with a specialist where appropriate.

Every effort has been made to trace the owners of copyright material used in this book, if any. The author and the publisher will be grateful for any omission brought to their notice for acknowledgement in the future editions of the book.

Published by : **Daya Publishing House®**
A Division of
Astral International Pvt. Ltd.
– ISO 9001:2015 Certified Company –
4736/23, Ansari Road, Darya Ganj
New Delhi-110 002
Ph. 011-43549197, 8130496929
E-mail: info@astralint.com
Website: www.astralint.com

Digitally Printed at : **Replika Press Pvt. Ltd.**

Foreword

Innovative mind-set is the hallmark of an evolved society. Quite clearly, societies that make innovation a way of life run far more efficiently than those that don't. Developing an ecosystem for consistent innovations, however, is a matter of policy persistence and therefore, calls for sustained investments. Streaming human resources toward creating an innovation ecosystem is as much a matter of public policy as it is, about the availability of competent manpower. Barring a few honourable exceptions, most of our public-funded higher academic institutions across the country are mired in incompetence and low productivity. The private sector's contribution too has been minimal because research is all about generating new knowledge and it takes a long time for the desired results to emerge – investors would be impatient. Barring the pharmaceutical sector, I cannot find many examples of private sector generating sufficient financial resources to do original research work especially in the primary domain of Agriculture and allied sectors. Most of the commercial seed producing companies being MNCs have begun their research elsewhere and extended that knowledge to India. That is one part of the story.

It is well established that the ROI on Agricultural Research is always the highest, the next best return being the investments that we make on rural roads. Therefore, it is necessary for a predominantly agrarian country to invest on appropriate agricultural research and develop the rural road infrastructure to enable the commodities to reach the markets. That said, I am often visited by a feeling that despite the heavy public investments that went into the agricultural research, the country's NARS has not been able to create an ecosystem for private research, or extension, or innovative entrepreneurship. The strength of the NARS, in a manner of speaking has become its weakness, because it appropriated a space that could have been jointly shared with farmers, innovators, and entrepreneurs. Farmer-led research therefore, remained an unexplored territory in India. Meanwhile the idea of need-based research has given way to scientist-centric or the institution-centric research. Only 11 per cent of the total research output of the NARS has ever been commercialized and how much of it has really reached the end-user farmers is unknown. That is a huge opportunity loss even when we grant that all research need not lead us to innovative products. I believe that organizations under the NARS should concern themselves with applied research - which they always do, as also promote and incubate innovations - which they rarely do.

Be that as it may, post 2014, a few changes have started taking place at the policy level that found some resonance among the youth, the innovators, and the current entrepreneurs. If the heart of economic growth is founded on innovation, making a product out of that innovation needs entrepreneurship. Country needs the innovators and entrepreneurs, and both will need an enabling framework. Until the financial system supports them to go to the next level, the innovator-entrepreneur combination will be helpless. A slew of programmes initiated by the national government to promote entrepreneurship seem to have started working. It may take years for the results to become tangible, but it is a worthwhile movement in the right direction.

It is in this context that the present addition to the current literature on innovation becomes valuable. Dr. R.M. Prasad, a reputed agricultural extension scientist has through a very diligent study of the world of innovation has come up with a document that will be useful to the students, serious academicians as also the practitioners. The Administrator always looks for frameworks and tools for implementing the programmes better and the academician always seeks to verify if there is indeed a theoretical foundation to the offered premises. I think Dr. Prasad has discovered this fine balance in his very well-researched book. The theoretical bases for innovation-led development is adequately presented while he also covered extension research relating to innovation to a good measure.

At a larger level, his effort would have been more rewarding if he had also scanned the innovation landscape of the country as to why certain states of the country are laggards while some others are early adopters. Why is such a vast difference between say the North East of India and Gujarat? What makes one state cross the chasm while the other finds it difficult? Can the micro theory relating to individuals be expanded to states within the country? How much of a role does the culture of a society play in the promotion of innovations? Are there some societies naturally more innovative than others? Do resource constraints make a difference to human mind with respect to innovations? What would be the impact of a financial ecosystem in promoting innovation? Also, why is that innovation in a sector like Information Technology is far more widespread relative to the primary sector? All these questions can wait for the second edition of this book as and when it happens.

At a personal level, I have known Dr. R.M. Prasad for several years now and know how much effort goes into anything that he does. The book reflects his effort, erudition and energy. This book will set a standard in the innovation literature of the country as being a unique and path-breaking contribution.

K. N. Kumar

Chairman,

Meghalaya Farmers' Empowerment Commission

Sri. K N Kumar with a Masters in Agriculture and Public Administration is a retired IAS Officer (1987 batch) presently serving as Chairman, Meghalaya Farmers' (Empowerment) Commission. He served as Agricultural Production Commissioner and Additional Chief Secretary, Government of Meghalaya. Sri. Kumar also worked as Chief Executive, National Fisheries Development Board, Hyderabad and as Deputy Director General, National Institute of Rural Development, Hyderabad. He was Deputy Director in Lal Bahadur Shastri National Academy, Mussoorie for four years.

Sri Kumar was instrumental in launching many commodity specific Missions in Meghalaya such as Aquaculture Mission, Mushroom Mission, Jackfruit Mission and Strawberry Mission. As Joint Secretary in the Ministry of Agriculture, GOI, he had conceptualized and drafted RKVY for accelerating the growth of the primary sector. He was also instrumental in the launch of placement linked skill development programmes in the country and PG programme in Rural Development Management in NIRD, Hyderabad. He received an Exceptional Contribution Award as a civil servant by G-Files Magazine in 2017, based on his innovative contributions in the field of agriculture and rural development.

Preface

Innovation is a term commonly used by all, but its meaning remains obscure. There are many related concepts used more or less as synonyms for innovation, such as modernization, brainchild, alteration, cutting edge, creation, invention, *etc*. The internet is often considered the ultimate example of innovation, which changed society in a significant way. However, internet has not emerged at one shot, but was built on the backs of many smaller, previous innovations.

Though innovation is most commonly associated with business and technology, it happens in many fields where people introduce changes. Innovations in the field of agriculture provide opportunity for farmers to increase productivity and manage natural resources efficiently. This helps to ensure long term viability and reduce the negative environmental impacts of production, such as pollutants and waste.

The students of Agriculture, particularly Agricultural Extension students are quite familiar with the name of E M Rogers, whose book on 'Diffusion of Innovations' is considered a classic book, which finds application in different disciplines. The diffusion of innovations dominated the studies on innovation in agriculture up to 1980s. It has emphasized technological innovations, but has not given sufficient importance to institutional innovations.

The theory of disruptive innovation emerging popular during the late 1990s have proved to be a powerful way of thinking about innovation driven growth. Disruptive innovations in agricultural sector have the potential to solve problems of scale, reduce information asymmetry and allow for supply chain disintermediation to make farming more profitable, inclusive and equitable. Clayton Christenson's books on disruptive innovations became more familiar to Agricultural Extension students during this period.

Sustainable development is a signifier which has been with us for more than two decades, during which time, there has been an unprecedented process of environmental degradation and deregulation, and a backtrack in the environmental front in the name of economy and growth. During this period, degrowth as a movement has attained significance, that aims at achieving a radical change in the metabolism of societies, towards a more frugal, sustainable and convivial lifestyle. Gerber (2020) has argued that critical agrarian studies (CAS) and degrowth can enrich each other. The agrarian question and the growth question need to be addressed in concert. At the core of degrowth is the climate crisis, which has profound influence on agriculture. The proponents of degrowth argue that to save Earth, we need to shrink global economic activity. Climate change should prompt a radical rethinking of economic growth, which is evident in various innovative ways of doing agriculture. Thus, innovation for degrowth is the need of the hour. The case studies of Pansera and Owen (2016) and Prasad and Dutta (2021) on degrowth are presented in this book in an abridged form for more clarity and understanding about degrowth.

The European Green Deal (EGD) is a new growth strategy unveiled by the European Commission in December, 2019 which aims for Europe to the first climate neutral continent by becoming a modern, resource efficient economy. The Farm to Fork (F2F) strategy at the heart of EGD addresses comprehensively the challenges of sustainable food systems and recognizes the inextricable links between healthy people, healthy societies and a healthy planet.

This book is an attempt to familiarize the readers with the concept of innovation and its various dimensions and the changing views on the concept of innovation in the field of agriculture. Beginning with the diffusion of technological innovations, the scenario gradually paved way for prominence of disruptive innovations and presently, degrowth innovations are gradually emerging as relevant in agriculture. As evident from the field, digital disruptive technologies in agriculture can lead to greater efficiencies that help countries advance agricultural practices and connect small scale producers to larger markets. However, the use of digital technologies in agriculture remains limited, often because of infrastructure, affordability, awareness, and regulatory issues.

The book is organized under ten chapter heads. The introductory chapter presents an overview of the definitions and meaning, innovation types and fields, importance of innovation in development and role of innovation in agriculture. The second chapter briefly covers the evolution and dynamics of innovation process. The common forms of innovation (social innovation, grassroots innovation, corporate innovation, jugaad and frugal innovation, local innovation and convivial innovation) are explained in this chapter. The third chapter on 'Innovation Ecosystem' explains the brief history of innovation policy in India. The key actors in the Indian innovation ecosystem are also explained. The innovation ecosystem in the agricultural sector is outlined in this chapter.

The chapter on 'Diffusion of Technological innovations in Agriculture' explains the concept and process of innovation diffusion, adoption and adopter categories, theories of innovation diffusion, attributes of innovation, barriers to adoption of innovations and issues related to innovation and diffusion studies. The fifth chapter on 'Disruptive innovations in Agriculture' presents an overview of disruptive innovations, types of disruptive technologies, smart innovations, NITI Aayog and disruptive innovations in Agriculture and disintermediation and uberization in agriculture.

'Innovation and Entrepreneurship in Agriculture' is the title of the sixth chapter which covers the sources of innovation in entrepreneurship, technology entrepreneurship, entrepreneurship in agriculture, agri business incubators and agri strartups. The chapter on 'Degrowth innovations in Agriculture' portrays the sources of degrowth theories, roots of degrowth movement, degrowth and technological innovations, successful cases of innovation in degrowth and studies on degrowth innovations in agriculture.

The chapter titled 'Innovation management and innovation intermediaries in Agriculture' explains the concept of innovation intermediaries, process, functions and typologies of innovation intermediaries and their role in the agricultural value chain. The chapter on 'Innovations in Science of delivery related to Extension services' explains the principles of science of delivery, models of extension service delivery, extension service as innovation brokers, institutional innovations in service delivery and the concepts of inclusive innovation and responsible innovation. The last chapter 'Epilogue' presents the important messages and outlines an innovative methodology for extension service delivery. The suggestions for innovation research in Agricultural Extension are also briefly presented.

The present publication has an important goal of highlighting the enormous potential and scope of innovations in agriculture and the need for blending insights from theories of diffusion, disruption and degrowth to generate new domains for extension teaching and research. I have tried to explain the various aspects related to innovation in a simple, non-technical language with a view to presenting the readers an easily readable version. The book is primarily targeted to students of Agriculture and allied fields, focused on students of Extension education. It is expected that the general public will also find this publication useful to gain an awareness about the different dimensions of innovations in the field of agriculture.

Since this publication is a rather lonely endeavour, it is quite possible that there could be many lapses and omissions considering the vast field of literature on innovation. Of course, I remain solely responsible for the errors and lapses, if any. I take this opportunity to request the readers to offer constructive criticism and valuable feedback for improvement of this publication.

I acknowledge the support and encouragement received from my family members, colleagues and students and this book is dedicated to them.

I am indebted to Sri. K N Kumar, IAS (Retd) who is a source of inspiration and motivation for many young minds. I am fortunate to get associated with him for short periods at NIRD and PR, Hyderabad and Meghalaya. He has written an articulated, thought provoking and scholarly Foreword for the book, which I strongly believe has added value and provided a clear vision.

I am thankful to Mr. Anil Mittal, Astral International Pvt Ltd, New Delhi for his co-operation and support for the publication.

R.M. Prasad

Contents

Acronyms

ABI	Agri Business Incubator
AC-ABC	Agri Clinic and Agri Business Centre
ACIC	Atal Community Innovation Centre
AGNIi	Accelerating Growth of New India's Innovations
AI	Artificial Intelligence
AIS	Agricultural Innovation System
AIM	Atal Innovation Mission
AIPSN	All India People Science Network
AKIS	Agricultural Knowledge and Information System
AKS	Agricultural Knowledge System
ARISE	Applied Research and Innovation for Small Enterprises
ARYA	Attracting Rural Youth in Agriculture
ASTI	Agricultural Science and Technology Institution
ATMA	Agriculture Technology Management Agency
AVC	Agricultural Value Chain
BoP	Bottom of the Pyramid
BIRAC	Biotechnology Industry Research Assistance Council
BVT	Bee Vector Technology

CDTA	Consortium of Researchers for Disruptive Technologies in Agriculture
CEIB	Centre of Excellence and Innovation in Biotechnology
CIIE	Centre for Innovation Incubation and Entrepreneurship
CIS	Community Innovation Survey
CoP	Community of Practice
CSA	Community Supported Agriculture
CSR	Corporate Social Responsibility
CSIR	Council for Scientific and Industrial Research
DARE	Department of Agricultural Research and Education
DAT	Disruptive Agricultural Technologies
DBT	Department of Biotechnology
DIME	Development Impact Evaluation Initiative
DOI	Diffusion of Innovation
DST	Department of Science and Technology
DSIR	Department of Science and Industrial Research
EU	European Union
FAO	Food and Agricultural Organisation
FBS	Farm Business School
FFS	Farmer Field School
FICCI	Federation of Indian Chamber of Commerce and Industry
GDI	Global Delivery Initiative
GFAR	Global Forum on Agricultural Research and Innovation
GII	Global Innovation Index
GNH	Gross National Happiness
GRI	Grassroot Rural Innovation
HLPE	High Level Panel of Experts on Food Security and Nutrition
HQCP	High Quality Cassava Peel
ICAR	Indian Council of Agricultural Research
ICRISAT	International Crops Research Institute for the Semi-Arid Tropics
ICT	Information and Communication Technology
IIGP	India Innovation Growth Programme
IDIA	International Development Innovation Alliance
IDEA	Indian Digital Ecosystem of Agriculture
INDSTA	Indian Science and Technology Archive of Research
IIO	Innovation Intermediary Organisation
IP	Intellectual Property

IPP	Innovation Policy Platform
IPM	Integrated Pest Management
IP & TM	Intellectual Property and Technology Management
IRTC	Integrated Rural Technology Centre
ISO	International Standards Organisation
ITC	Indian Tobacco Company
ITMU	Institute Technology Management Unit
KET	Key Enabling Technologies
KIA	Knowledge Initiative in Agriculture
KIBS	Knowledge Intensive Business Services
KPI	Key Performance Indicator
KSSP	Kerala Sastra Sahitya Parishad
MANAGE	National Institute of Agricultural Extension Management
MFI	Micro Finance Institution
MSME	Micro, Small and Medium Enterprises
MVIF	Micro Venture Innovation Fund
MVP	Minimal Viable Product
NAARM	National Academy of Agricultural Research and Management
NABARD	National Bank for Agriculture and Rural Development
NAFPO	National Association of Farmer Producer Organisations
NAIF	National Agricultural Innovation Fund
NAIP	National Agricultural Innovation Project
NASSCOM	National Association of Software and Service Companies
NATP	National Agricultural Technology Project
NeGD	National E-Governance Division
NESTI	National Experts on Science and Technology Indicators
NICRA	National Innovation on Climate Resilient Agriculture
NIDHI	National Initiative for Developing and Harnessing Innovation
NIF	National Innovation Foundation
NMITLI	New Millennium Indian Technology Leadership Initiative
OECD	Organisation for Economic Co-operation and Development
OSAT	Open Source Appropriate Technology
PGS	Participatory Guarantee Scheme
PKVY	Paramparagat Krishi Vikas Yojana
PoC	Proof of Concept
PRISM	Promoting Innovation in Individuals, Startups ad MSMEs
QR code	Quick Response code

RCT	Resource Conservation Technologies
RFID	Radio Frequency Identification
R & D	Research and Development
RKVY	Rashtriya Krishi Vikas Yojana
RRI	Responsible Research and Innovation
RTK	Real Time Kinematic
SBIRI	Small Business Innovation Research Initiative
SDG	Sustainable Development Goal
SHG	Self Help Group
SI	Social Innovation
SIMRA	Social Innovation in Marginalised Rural Areas
SGSB	Stanford Graduate School of Business
SRI	System of Rice Intensification
STI	Science, Technology and Innovation
STIP	Science, Technology and Innovation Policy
TAM	Technology Acceleration Model
TAP	Tropical Agricultural Platform
TBI	Technology Business Incubator
TIFAC	Technology Information Forecasting and Assistance Council
UNIDO	United Nations Industrial Development Organisation
USDA	United States Department of Agriculture
UTMT	Under the Mango Tree
VCA	Value Chain Analysis
WIPO	World Intellectual Property Organisation
ZBNF	Zero Budget Natural Farming
ZTMC	Zonal Technology Management Centre

Chapter 1
Introduction

It is an accepted fact that the success of any individual or organisation is dependent on innovation. Innovation allows individuals and organisations to stay relevant in the competitive market and plays an important role in economic growth of a nation.

Over the last decades, innovation has become a significant means to combat critical social risks and threats. The Policy brief of OECD (2021) discusses the effects that the COVID-19 crisis could have on the future of science, technology and innovation (STI) and its policies. The accelerated adoption of digital tools and techniques is highlighted. The crisis could also spur experimentation with new tools, policy approaches and governance models.

Innovation is responsible for resolving various problems in the society in a sustainable and efficient way, usually with new technology. The new technologies, products and services simultaneously meet a social need and lead to improved capabilities and better use of assets and resources. From the society's perspective, the fundamental outcomes of innovation are economic growth, increased well-being, improved communication and accessibility and environmental sustainability.

Innovation also increases the chances of organisations to react to various challenges and discover new opportunities. It can also help foster competitive advantage as it allows to develop better products and services for the customers.

1.1. Innovation Defined

The word "innovation" is derived from the Latin verb 'innovare', which means to renew. Innovation denotes the means to improve or to replace something, a process, a product, or a service. Innovation is the practical implementation of ideas that result in the introduction of new goods or services or improvement in offering goods or services.

Webster's defines innovation as *"the introduction of something new."*

According to Peter Drucker (1985), *"Innovation is the specific instrument of entrepreneurship... the act that endows resources with a new capacity to create wealth"*. The concept of endowing resources with a new capacity to create value is interesting.

Scott Berkun (2010) remarked, *"Innovation is significant positive change."* He adds the notion of "significance." This definition focuses on the outcome rather than the process, and while the word "change" implies novelty, it is probably too broad.

Tim Kastelle (2012) views innovation as *"not just having an idea, but executing it so that it creates value."* He incorporates three key concepts: an idea, execution, and value creation.

Innovation is the implementation of something new or improved (whether technology or otherwise) in products (goods or services), processes, marketing or organizational methods. In other words, it means applying ideas, knowledge or practices new to a particular context for creating positive change that will provide a way to meet needs, take on challenges or seize opportunities. Such novelties and useful changes could be substantial (a significant change or improvement) or cumulative (small changes that together produce a significant improvement) (OECD, 2012)

Innovation is the process by which individuals or organisations master and implement the design and production of goods and services that are new to them, irrespective of whether they are new to their competitors, their country or the world (World Bank, 2006).

ISO TC 279 in the standard ISO 56000:2020 defines innovation as "a new or changed entity realizing or redistributing value".

It can be inferred from the above that innovation is a complex process and defining innovation is not easy. The genesis of innovation derives from a wide range of sources and its development involves various stages, often involving considerable investment.

1.2 Meaning of Innovation

The definition for innovation varies widely. Skillicorn (2016) after synthesising various meanings furnished by the innovators, explained

Box 1.1

Skillicorn (2016) based on interviews with 15 innovators has provided the meaning of innovation, as below:

- ☆ Application of ideas that are novel and useful
- ☆ Innovation is about staying relevant
- ☆ Great idea, executed brilliantly, and communicated well
- ☆ Feasible, relevant offering with a viable business model perceived as new and adopted by customers
- ☆ Introduction of new products and services that add value to an organization
- ☆ As long as it includes "new" and addresses customer needs, any variation goes as innovation
- ☆ Fundamental way companies bring constant value to their customers
- ☆ Work that delivers new goodness to customers in new markets and radically improves the profitability equation
- ☆ Implementation of something new
- ☆ Implementation of creative ideas to generate value
- ☆ Anything new, useful, and surprising

Source: Skillicorn, Nick (2016) Infographic: 15 Experts on what innovation actually means. http://www.inc.com>nick-skillicorn

innovation as "Turning an idea into a solution that adds value from a customer's perspective". Value creation is considered an essential feature of innovation.

Based on analysis of various definitions, innovation can be summarised as:

- ☆ The process of creating value by applying novel solutions to meaningful problems.
- ☆ The practical application of ideas that result in different new types of offerings, like products, services, processes, and business models, intending to improve or disrupt existing applications or creating new solutions.
- ☆ The process by which a domain, a product, or a service is renewed and brought up to date by applying new processes, introducing new techniques, or establishing successful ideas to create new value. The creation of value is a defining characteristic of innovation.
- ☆ Being creative and original in the work and thinking. Every brand, person, and product wants to be perceived as innovative, but what separates actual innovation from everything else is the idea of improvement.

Innovations vary along different dimensions. Some of the dimensions are types, fields, and novelty of innovation.

1.3. Types of Innovation

The Oslo Manual, an international reference guide from the Organisation for Economic Co-operation and Development (OECD) for collecting and using data on innovation, defined the concept of innovation as: *"a new or improved product or process (or a combination thereof) that differs significantly from the unit's previous products or processes and that has been made available to potential users (product) or brought into use by the unit (process)."*

The Oslo Manual distinguishes between innovation as an outcome (an innovation) and the activities by which innovations come about (innovation activities).

According to the most recent edition (4th edition, 2018) of the Oslo Manual, there are four main types of innovation:

Organizational Innovation

It refers to the development of a new organizational strategy that will somehow change a company's business practices, as well as the way its workplace is organized and its relationship with external stakeholders. A recent example is the 'Work from Home' policy adopted by the companies to overcome COVID-19 crisis.

Box 1.2

The Oslo Manual is an essential worldwide tool for statisticians and policy makers concerned with innovation. It includes definitions of basic concepts, data collection guidelines, and classifications for compiling innovation statistics. The 2018 edition contains improved guidance reflecting evolving user interests and accumulated practical experience. It includes new material dedicated to supporting the measurement of innovation outside the business sector, understanding the internal and external drivers of business innovation as well as a firm's most important innovation, and facilitating better use of innovation data for statistics and analysis.

The Oslo Manual is based on the experience gained from collecting innovation statistics in both OECD and EU countries and non-member countries. It is the result of collective work of National Experts on Science and Technology Indictors (NESTI) and Eurostat's Community Innovation Survey (CIS) Task Force, involving experts from nearly 45 countries and international organisations.

Source: Oslo Manual 2018 Guidelines for collecting, reporting and using data on innovation http://oe.cd/oslomanual

Process Innovation

It is about implementing a new or improved production or delivery approach, including changes in operational methods, the techniques used and the equipment or software. Examples include use of cloud computing, block chain technology, *etc.*, which find application in agriculture.

Product Innovation

It is the introduction of a new or improved good or service. These inventions or changes may have to do with improving technical specifications, the materials or the software used or even advancing on UX (user experience). Use of electric vehicles is an example for product innovation for addressing environmental problems.

Marketing Innovation

It means developing a new marketing strategy that produces changes in, for instance, the way a product is designed or packed, or even other decisions regarding price or promotion. Examples include use of edible cups and plates replacing environmentally damaging plastic cups and plates.

There is also another type of innovation called **'service innovation'**, which refers to the creation and introduction of new services for customers and other stakeholders.

Another type of innovation is also added, referred to as **Eco-innovation**, which is an innovation resulting in significant progress towards the 2030 sustainable development goals (SDGs) established by the UN. In practice, it means reducing the impacts of our production modes on the environment, improving nature's resilience to environmental stresses, or achieving a larger efficient and responsible use of natural resources. The concept of degrowth has its roots in eco-innovations.

1.4. Doblin's 10 Types of Innovation

The approach to innovation that looks at innovation through the lens of ten dimensions was proposed by Doblin, a global innovation agency founded in 1981 by Jay Doblin and Larry Keeley. The ten types of innovation theory is the result of almost 40 years of studies and research on over 2000 successful companies, such as Amazon, IBM, and Ford. The researchers at Doblin aimed to identify patterns in companies which have successfully innovated in the past, as well as the reasons for types of initiatives that have failed.

The ten types of innovation identified by Doblin fit in three categories- Configuration, Offer and Experience.

A. Configuration

This category contains all the elements that go inside a business and the systems which make the foundation of business.

1. **Profit Model Innovation** deals with how the organization makes money, innovating about where the revenue is coming from, and what can be done to generate more revenue. Netflix is a classic example for profit mode, the way in which it managed to radically transform the movie rental business, at first, into a DVD-delivery one and then, into the streaming industry. The Swiss agricultural company Syngenta's innovative way to package and distribute crop protection chemicals to small holders in Kenya (uwezo) along with investment in capacity building to teach the farmers how to use the product, built an ecosystem that included retail distributors, which resulted in both profit maximisation and small holder productivity.
2. **Network Innovation** looks the way in which the company deals with the relationship with the stakeholders and how the company approaches the strategies behind the partnerships. It is a form of open innovation, which is the result of sharing of knowledge and expertise between multiple companies. Supply chain of a company is an example for network innovation.

   Open innovation is an example for network innovation model wherein companies utilise each other's resources and technology and also sharing risks and awards with each other. It is basically an alternative to the conventional method (closed innovation) of doing innovation, where information has to stay within pre-set confines.
3. **Structure innovation**–To innovate at a structural level, the organisation must look at the tasks and processes which stand at the base of the organization, and deals with the assets of the company, both tangible and intangible and how to arrange them to give the best results and performance. General Electric's (GE) eco-imagination project allows embracing crowd-sourced innovation and collaborative problem solving to address customer needs more effectively and more efficiently. Eco-imagination refers to promoting renewable energy sources and developing new technologies to reduce carbon emissions. This strategy enables GE to compete in countries with increasingly strict regulatory environments and take advantage of carbon tax credits.

   An example of structure innovation comes from Google, with its 20 per cent Project, where employees can work, for 20 per cent of their time, on personal projects that can also benefit the company. Whole Foods, certified organic grocer is another example. Whole Foods market has a global hierarchy with four main tiers- global headquarters, regional offices, facilities and stores. This hierarchy supports corporate monitoring and control over the entire enterprise.

4. **Process innovation** is when the processes inside the company are improved for their efficiency either to reduce production time or to reduce certain costs.

   McDonald's is one of the most famous process innovation examples, which have the kitchen process that made the company famous. Tata Group's Computational Research Laboratories (CRL) developed a parallel processing library technology, that has been able to develop the fourth fastest computer in the world.

B. Offer

This category is about what the company can offer the clients, in terms of products, services, experience and value.

5. **Product Performance Innovation**–focuses on the value and the features and quality of the products. This is done by developing new products or by improving products that already exist.

   An example of product performance innovation is Gorilla Glass, made by Corning, which is the result of continuous testing and improvement. The Jaipur Foot, a hand-made artificial foot, is another example of a successful product performance innovation.

6. **Product System Innovation**–It is about the company's offer which satisfies the customer. Along with the main product, other complementary offers that can be added to satisfy the customers' needs better, and also bring profit for the company are carefully chosen and offered to the customers. A classic example is Nike +, a kit and an app that brings together multiple Nike products and offers.

C. Experience

This category is focused on the user, on his or her experience when interacting with the products, services and all other aspects of the company.

7. **Service Innovation**–deals with the 'relation' between the customers and the product, especially the improvement of the product in terms of usability. This category of innovation talks about how the business provides 'customer service' to its clients which helps in using the product.

   Ninjacart, India's largest fresh produces supply chain company is a good example for service innovation in agriculture. The company sources fresh produces from farmers and deliver them to customers within 12 hours.

Box 1.3

Ninjacart is India's largest Fresh Produce Supply Chain Company. They are pioneers in solving one of the toughest supply chain problems of the world by leveraging innovative technology. Ninjacart source fresh produce from farmers and deliver them to businesses within 12 hours.

The company connect producers of food directly with retailers, restaurants, and service providers using in-house applications that drive end to end operations.

The company is focused on making the Ninjacart innovation more accessible to the most fragmented parts of society. Ninjacart intend to leverage their strengths and resources to innovate for new product categories and customer segments while solving complex supply chain problems.

Source: https://www.ninjacart.in

8. **Channel innovation**- is about all the channels where the customers and potential customers can find out about the company and its offering, as well as the way in which the company communicates and maintains that relationship.

   An interesting example of channel innovation is from the New York Public Library which created the Insta Novels – a very efficient way to connect with a new generation of readers, offering them the right content on the right platform. Tesco Korea's virtual store is another good example. The virtual stores are located at platforms so that the commuters can shop via their smartphones using QR code.

9. **Brand innovation**- is about the company's brand, which makes sure that the potential customers place the brand before those of the competitors. A brand image goes a long way in sustaining a business.

   A great example of Brand Innovation is 'Amul- The Taste of India'. Microsoft is another example which uses the power of storytelling to make its content compelling through Microsoft stories.

10. **Customer Engagement Innovation** deals with what the customers think about the product or service and how the company can interact with them in a more efficient way. This is done by identifying the needs of the consumers.

    A great Customer Engagement Innovation example is Apple. Whenever Apple launches new products, it manages to keep a solid loyal customer base. Disney is another example, which sells magic in vacation packages. Disney has learned the trick of customer engagement through its various platforms.

Box 1.4

Innovation is essentially about weaving together 'knowledge spaghetti', combining different knowledge strands- technical, market, financial, legal, *etc.*- to create value. Innovation happens through a process often referred to as 'the spaghetti model', where it is dependent on a seemingly chaotic and messy blend of the right elements at the right time. One of the most important of these elements are people.

Innovation is introduced as a process with a defined structure, implying a process with clear stages. In reality, innovation is complex. The way in which knowledge flows around an innovation project are complex, interactive, woven together in a kind of social spaghetti where different people talk to each other in different ways, about different things. The interaction is mainly about knowledge flow. The bigger the innovation project, the more complex the innovation networks become. Not only do the actors in the networks increase, but also the platforms which support these relationships. The relationships can become more virtual, global, enabled by technologies of the internet, broadband, mobile communication and shared computer networks.

Source: https://www.coursehero.com>file

1.5. Fields of Innovation

Innovation can be in different forms and outcomes. When discussing about innovation, most people think of only new products, while a wide array of different innovation outcomes are possible. The various innovation outcomes are given below (Talin, 2021).

1. Product and Product Performance Innovation

Either a new product is developed or the performance of an existing product is improved. This kind of innovation is the most common in the business world. New versions of cellular phones and computers introduced in the market every year with more capabilities and uses and lower prices is an example for product innovation.

2. Technology Innovation

New technologies can be the basis for many other innovations. Technology has considerably reduced transaction costs. The best example is the Internet, which itself is an innovation that also lead to other innovations in various fields. E-commerce, e-trainings, *etc.* are innovative technology applications.

3. Business Model Innovation

Many of the most successful companies in the world have managed to innovate their business model. Using different channels, technologies and new markets can lead to new possible business models which can create, deliver and capture customer value. Digital ecosystems are a well-known example of innovation using several technologies and creating an altogether new type of business.

Alibaba group is the largest online retail company in the world, which has no inventory. Alibaba's business model is to connect buyers with sellers.

4. Organizational Innovation

Managing and sharing resources in a new way can also be an innovation. This way it's possible to use resources and assets in a completely new way.

Starbucks Corporation (Starbucks Coffee Company) uses its organised structure to facilitate business development in the global coffee industry.

5. Process Innovation

Process innovation includes changes in the technology and equipment used in the design, development and manufacture of products or services. Innovation in the processes can improve the efficiency or effectiveness of existing methods. Possible process innovations involve production, delivery, or customer interaction.

The blockchain technology in agriculture enables the traceability of information in the food supply chain, which helps improve food safety. It also provides a secure way of storing and managing data, which facilitates the development and use of data-driven innovations for smart farming and smart index-based agriculture insurance.

6. Marketing/Sales – New Channel Innovation

Here, new methods are used to capture and hold attention from customers, either through the use of innovative marketing/sales concepts or the use of new channels for customer acquisition/sales.

There is an enormous market of customers for farm microfinance products, and only a handful of institutions are serving them. To meet even a fraction of this demand, more of microfinance institutions (MFIs) are needed to develop new products and deliver them to their customer bases at scale. There is potential for MFIs in the agricultural sector, which can be tapped.

7. Network Innovation

By connecting different groups and stakeholders it might be possible to create extra value. This type of innovation is very common due to the use of ICT services.

LoRaWAN is a blueprint for allowing devices with limited power capacity to connect to wireless networks covering a wide geographic area.

8. Customer Engagement/Retention Innovation

Innovative concepts try to increase the engagement of customers and keep the retention up. The goal is to have innovative models to keep the customers "locked-in" or engaged.

Amazon focuses on the customer experience, while Nestle monitors and improves services based on customer feedback.

1.6. Novelty of Innovation

Based on the novelty of technology and market, four types of innovation are identified. They are:

1. Incremental Innovation (Existing Technology, Existing Market)

This is one of the forms of innovation most commonly observed. It uses existing technologies within an existing market. Incremental innovation focuses on marginal improvements to what already exists. The goal is to improve an existing offering by adding new features, changes in the design, *etc.* This form of innovation is the vital pillar behind the longevity of many of the oldest and recognised brands of the world.

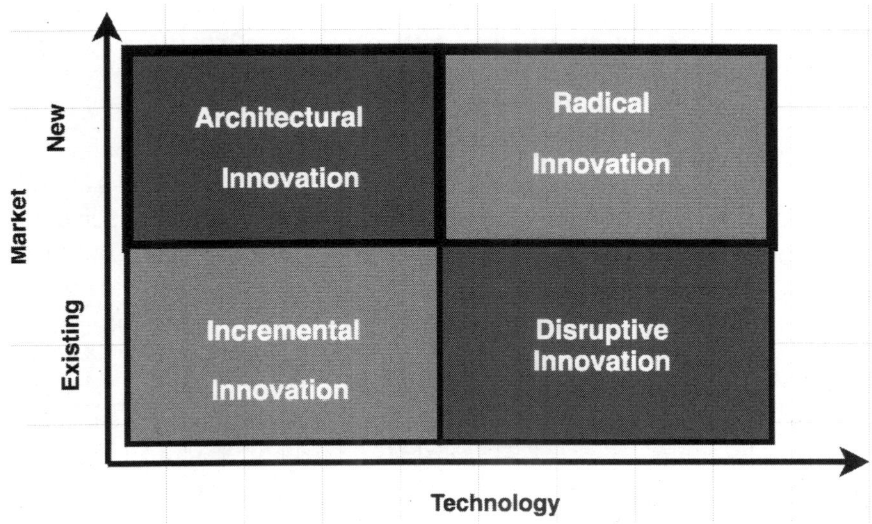

Figure 1: Types of Innovation.

The best example for incremental innovation is the Smartphone market where mostly innovation is only updating the hardware, improving the design, or adding some additional features/cameras/sensors, *etc.* Another example is Amul, which has innovated introducing new flavours and new dairy products.

2. Disruptive Innovation (New Technology, Existing Market)

It is the process by which a small, less known company, usually with less of resources, moves upmarket and competes with large established businesses. Disruptive innovation is mostly associated with applying new technologies, processes, or disruptive business models to existing industries. Sometimes new technologies and business models seem, especially in the beginning, inferior to the existing solutions but after some iterations, they surpass the existing models and take over the market due to efficiency and/or efficacy advantages.

Amazon used Internet-Technologies to disrupt the existing industry for book-shops. They had the existing market for books but changed the way it was sold, delivered and experienced due to the use of disruptive technologies. Another example is the iPhone, where existing technologies in the market (Phones with buttons, keypads, *etc.*) were replaced with touch-interface-centred devices combined with intuitive user interfaces. Use of drones in agriculture for application of plant protection chemicals is another example.

3. Architectural Innovation (Existing Technology, New Market)

Architectural innovation refers to changing the overall design of a product based on customer demand. It is something we see with tech giants like Amazon, Google, and many others at the moment. They take their domain expertise, technology, and skills and apply them to a different market. This way they can open up new markets and expand their customer base.

The digital ecosystem orchestrators like Amazon and Alibaba use this innovation strategy to enter new markets. They use existing expertise in building apps, platforms, and their existing customer base to offer new services and products for different markets. A recent example for this is Amazon entering the medical care field. Likewise, Agricultural hub is an architectural innovation that can serve as the central facility for the farmers and the consumers.

4. Radical Innovation (New Technology, New Market)

Radical innovation involves the creation of technologies, services, and business models that open up entirely new markets. It is a transformative business model that seeks to completely replace an existing industry or create a new industry.

An example of radical innovation is John Deere's innovation, who took notice of emergent sensory technology and added sensors to its tractors. Farmers could track and monitor the crop yields via sensors, which revolutionalised the farming sector.

1.7. Innovation Process

The innovation process describes the path of translating new and/or existing knowledge into marketable solutions. Jan Spruijt (2016) had developed an innovation process model, which is a simple design that makes it easy to bridge the gap between a formal process and the tools available.

The following eight types of innovation process are identified.

- ☆ Marketing and Branding: Innovating the customer experience
- ☆ Ideation: Innovating the product idea and concept
- ☆ Technology: Innovating the product functionality
- ☆ Co-creation: Innovating the customer involvement
- ☆ Social innovation: Innovating the corporate culture
- ☆ Entrepreneurship: Innovating through entrepreneurial thinking
- ☆ Open innovation: Innovating with stakeholders
- ☆ Business model innovation: Innovating the purpose and strategy

1.8. Four P's of Innovation Strategy

John Bessant and Joe Tidd (2007) summarised innovation strategy into 4 P's, as below:

1. **Paradigm innovation:** shifting of an underlying mental model about a product via a changed version. *e.g.,* iphone
2. **Process innovation:** focuses around change in operational processes, which can impact the efficiency and profitability of a business. *e.g.,* on line marketing
3. **Position innovation:** positioning innovation is about repositioning, taking a product or service and offering it in a new market or with a new introduction. *e.g.,* Starbucks offers one of the classic examples of brand repositioning. in coffee business.
4. **Product innovation:** this is the most commonly thought of innovation category. *e.g.,* New versions of PCs.

1.9. Eight P's of Innovation

Joost Korver, Chief Entrepreneur and Head of Growth at Gasbot, an Australian IoT startup has proposed 8 P's of innovation as below:

1. Processes (innovation strategy, governance, innovation methodology, team structure)
2. People (finding the right people, training)
3. Projects (methodology: lean start up, design sprints)
4. Problems (start with real problems and not pet projects from a few)
5. Priorities (Innovation thesis, aligned with strategy)
6. Progress (Measure progress!)

7. Partnerships (collaborations and external inspiration/trend watching *etc.*)
8. Places (create physical and mental space to be creative)

A basis of starting any innovation journey is the triangle of people, processes and projects. A good structure and processes are needed, as well as the right people with the right mindset and attitude.

The best way to start learning about innovation is by actually doing it. Get started with some innovation projects and ensure that the right support is obtained to adopt the right principles and methodologies.

An innovation journey usually starts when an organization is in a tight spot and under pressure. It is not sufficient to just do some innovation projects, it is also important to have the right focus and ensure your innovation activities are aligned with the overall company strategy.

Establishing meaningful connections and partnerships is the next important area. Creating spaces where creativity is facilitated is powerful and contagious.

People perceive different things when they hear the word 'innovation'. In today's world everyone talks about digital transformation, disruptive innovation, business innovation, *etc.* without properly understanding the meaning and context. The core innovation definitions along with related processes, methodologies, practices, and innovation artifacts are given as **Annexure-1**

1.10. Innovation and Development

Innovation is essential for development as it is a driver of economic growth. In economic terms, innovation describes the development and application of ideas and technologies that improve goods and services or make the process of production more efficient. Innovation can make a difference in addressing urgent developmental challenges such as providing access to drinking water, eradicating neglected diseases or reducing hunger. The build-up of innovation capacities has played a central role in the growth dynamics of successful developing countries.

Innovation is important to the advancement of society as it solves various kinds of social problems and enhances society's capacity to act. It is responsible for resolving collective problems in a sustainable and efficient way, usually with new technology. From the society's perspective, the fundamental outcomes of innovation are economic growth, increased well-being and communication, educational accessibility and environmental sustainability. Technological innovation is considered a major source of economic growth.

Many growth-enhancing innovations also address social challenges. For example, poverty-related effects can substantially influence opportunities for engaging in entrepreneurial activities. While innovation is important at all stages of development, different types of innovation play different roles at various stages. In earlier stages, incremental innovation is often associated with the adoption of foreign technology, and social innovation can improve the effectiveness of business and public services. High-technology R&D-based innovation matters at later stages of development, when it is both a factor of competitiveness and of learning.

Information and communication technologies (ICTs) offer many opportunities for innovation. They drastically reduce the cost of accessing or diffusing information. ICTs have also facilitated the emergence of global value chains, considered as engine for industrial and technological development in a number of countries, allowing technology transfers from multinational firms. Global value chains offer a potentially different framework for development. "Dematerialised" innovation (*e.g.* product design) is more relevant today for countries' positioning in global value chains.

Innovation essentially paves the way to creatively improve the standard of living and satisfy the needs of consumers. The earliest literature on innovation can be traced back to Adam Smith's "An Inquiry into the Nature and Causes of the Wealth of Nations", which explores the role played by innovators in leading technological improvements in factories and society. However, it was Joseph Schumpeter who accorded innovation the centrality in its role of pushing economic growth and industry transformation.

The innovation ecosystem runs on both material capital, such as funds, equipment, facilities, as well as human capital, which makes up the various institutional actors and entities participating in the ecosystem. The innovation ecosystem is run by two parallel economies, commercial and research. While fundamental research is the key driver of research economy, commercial economy is primarily driven by the marketplace.

1.11. Challenges of Innovation

According to Michael Porter (1980), the world's foremost thinker on modern strategy, innovation is a key component in the characterization of development strategy. Porter introduces one of the most powerful competitive tools: his three generic strategies, *viz.*, lowest cost, differentiation, and focus, which bring structure to the task of strategic positioning. He shows how competitive advantage can be defined in terms of relative cost and relative prices, thus linking it directly to profitability, and presents a new perspective on how profit is created and divided.

An innovation can benefit more people and companies than just the innovating firm. If the firm cannot charge all the beneficiaries of its innovation, then there is a problem of matching incentives to the value of the activity, which may lead to an undersupply of innovations. The possibility that the market system, guided by the independent actions of private firms, will not lead to the optimal outcome is called "market failure."

There are two possibilities of market failure by innovations. The first is that new knowledge, which is created during the innovation process, is what economists term a public good and such goods tend to be underprovided by the private market. The second is that innovation can create positive externalities in the form of spill-over benefits to customers and other firms and these cannot be captured as revenue by innovating firms, again leading to under-provision of innovation.

Many aspects of the innovation process are subject to market failures and the existence of the intellectual property (IP) regime is an attempt to remedy some of the problems.

1.12. Innovations in Agriculture

Innovation in agriculture is the central driving force for achieving a world free from hunger and malnutrition. Innovation in agriculture is particularly relevant for addressing socioeconomic challenges and fostering growth at the same time. There is evidence that agricultural R&D has a greater impact on poverty reduction than most other public investments. Innovation is generally perceived as related to technologies. In fact, agricultural innovation is much broader, which is the process whereby individuals and organisations bring new or existing products, processes or ways of organisation into use for the first time in a specific context, to increase effectiveness, competitiveness and resilience with the goal of solving a problem faced by the farmers. FAO plays a key role in promoting the importance of innovation in agriculture to increase food security, sustained development and promote rural development.

Innovation in agriculture cuts across all dimensions of the production cycle and along the entire value chain - from crop, fishery, forestry or livestock production to the management of inputs to market access. FAO harnesses the power of digital technologies to pilot, accelerate and scale innovative ideas with high potential for impact in food and agriculture, transforming digital solutions and services into global public goods. It aims to explore the responsible application and adoption of existing and frontier technologies, design and scale new services, tools and approaches to empower rural households and inspire youth entrepreneurship in food and agriculture.

1.13. Organization of the Book

Agricultural development enables agriculture and farmers to adapt rapidly when challenges occur and to respond readily when opportunities arise. In the agricultural sector, innovation is considered vital for sustainable economic, social, and ecological development.

Innovations in agriculture provide both opportunities and challenges. Responding to the challenges and opportunities calls for not just technological innovation, but innovations in how societies understand and value their food systems; innovations in the market; incentives that can drive agricultural and food systems toward equity and sustainability; and innovations in the way farmers, government, business and civil society work together to govern good systems and natural resource management.

The innovations in the field of agriculture can be analysed in three phases. The book addresses innovations in relation to the three phases- technological innovations, disruptive innovations and innovations for degrowth.

In the first phase, technological innovations have resulted in increased production and productivity of crops, livestock and fisheries. In this respect, the diffusion process of technological innovations is perhaps one of the most widely researched and documented social phenomena. A number of researches have been conducted in the field of diffusion and adoption of agricultural innovations providing invaluable knowledge base regarding the processes and factors that directly or indirectly affect them.

In the second phase, disruptive innovations in agriculture, particularly related to digital technologies have resulted in more precision and efficiency in production. Disruption refers to innovation that creates new market and value network and displaces existing and established products and procedures. The innovations in entrepreneurship in agriculture related to agricultural enterprises and startups also have emerged during this phase. NITI Aayog, Government of India accords high importance in promoting disruptive technologies as part of their effort to enhance the income of farmers.

In the third and recent phase, the focus of innovation is on degrowth. The point of degrowth is not to have 'less of the same', but to organise appropriation, extraction, production, distribution, consumption and waste differently. Degrowth is not anti-technology, per se. It refers to planned reduction of energy and resources use designed to bring the economy back into balance with the living world. Degrowth theories dismantle the automatic association of 'growth' with 'improvement'. Natural farming promoted by the Government of India can be considered a degrowth innovation concept.

Besides covering the three phases of innovation related to agriculture, the book also covers the innovation ecosystem in agriculture, innovation intermediaries and innovations in service delivery to farmers. Innovations in service delivery covers both institutional innovations and also process innovation.

REFERENCES

Berkun, Scott (2010) The Myths of Innovation, O' Reilly

Bessant, John and Tidd, Joe (2007) Innovation and Entrepreneurship, Wiley

Drucker, P F (1985) Innovation and Entrepreneurship, UK: Pan Business Management

ISO-TC 279 Innovation Management. http://committee.iso.org>home>tc279

Jan Spruijt (2016) 8 Types of Innovation Process. http://openinnovation.eu/14.09.2016/8_types_of_innovation_processes

Kastelle, Tim (2012) What is Innovation? http://timkastelle.org>blog>2012/09,what_is_innovation

Korver, J (2020) A Successful Approach to Corporate Innovation: 'The 8 Ps of Innovation

https://www.linkedin.com › pulse › successful-approach-corporate-innovation

OECD (2005) Innovation for Development, http://www.oecd.org>innovation>innovation_for_development

OECD (2021) How will COVID-19 reshape science, technology and innovation? OECD Policy Responses to Coronavirus, 23 June, 2021.

Porter, Michael (1980) Competitive Strategy, Free Press, New York

Talin, Benjamin (2021) Innovation. http://www.morethandigital.info/en/innovation_definition_innovation_types_and_meaning

World Bank (2006) Enhancing Agricultural Innovation: How to go beyond the strengthening of Research Systems, Washington D C, World Bank

Chapter 2

Evolution and Dynamics of Innovation Process

2.1. Innovation as a Process

The process of innovation evolves in the same way as humans evolve. Looking at humanity's slow rise from prehistory, three waves or evolutionary acceleration can be identified. The first was the transition from hunting to the development of the agricultural age, propelled by handmade tools. The second was the industrial revolution, propelled by the invention of automation. The third is the form of information revolution using IT tools.

Innovation also evolved in a similar fashion. The processes and technologies that comprise the art and science of innovation have progressed in three oddly similar phases. During the "before computers" (BC) era, innovators were equivalent to hunter gatherers of ideas. But in the "After Digital" (AD) era, the invention of computers changed the whole outlook (Moses, 2012).

The first wave of innovation evolution came with the invention of early innovation tools such as mind-mappers, idea catchers, *etc.* The second wave of evolution started with the deployment of innovation pipelines and stage gate technologies that allow the production of ideas to be industrialized and automated. This was the era of the "idea factory". Third wave innovation seeks not only to simplify and automate the production and processing of ideas, it aims to enable inventors to create fundamentally better ideas. The use of AI and robotics assist humans in the process of refining ideas.

Dynamics refers to movement, change and flux. Analysis of how innovations enter and transform enterprises reveals several strategies for mastering innovation as a creative force. Technological change is a key factor as both a creative force in the growth of enterprises and also as a deterrent in making enterprises vulnerable to competition.

Besides technology, innovation has other dimensions also such as social, organizational and institutional, which can take different forms. Innovation is not a top-down or controlled process. Rather, innovations often emerge from the bottom-up, in an evolutionary way.

Innovation process can be considered as a knowledge transformational process using three major activities: (a) Activity of knowledge sourcing, (b) Activity of knowledge transformation into a physical innovation (new product or process) and (c) A final stage of activity is dedicated to the exploitation of innovation through commercialization.

2.2. Design Thinking as a Strategy for Innovation

Design thinking can be described as a discipline that uses the designer's sensibility and methods to match people's needs with what is technologically feasible and what a viable business strategy can convert into customer value and market opportunity. (Brown, 2008).

When design principles are applied to strategy and innovation, the success rate for innovation dramatically improves. Designing is more than creating products and services; it can be applied to systems, procedures, protocols, and customer experiences.

Design is transforming the way leading companies create value. The focus of innovation has shifted from being engineering-driven to design-driven, from product-centric to customer-centric, and from marketing-focused to user-experience-focused.

Design thinking draws on logic, imagination, intuition and systemic reasoning to explore the possibilities of what could be and to create desired outcomes that benefit the end user (the customer). It provides innovators with a framework for addressing complex human-centred challenges and making the best possible decisions concerning:

- Redefining value
- Re-inventing business models
- Shifting markets and behaviours
- Organisational culture change

- ☆ Complex societal challenges such as health, education, food, water and climate change
- ☆ Problems affecting diverse stakeholders and multiple systems

Design thinking is a human-centred approach employed to better understand customer needs and design innovative products and services that can fuel business growth. This approach to innovating is focused on understanding the end user's needs, quick prototyping, and developing new ideas. Design thinking provides tools and techniques that ensure the firm is solving the right customer problem before rushing to a solution.

2.3. Generic Framework for Designing the Innovation Process

Though there is no standard or typical innovation process, there is a generic framework common for almost all sectors and companies. It can be used as a basis for designing the individual innovation process. Taking an innovation from the first spark through to a finished product involves seven steps:

1. **Strategic direction:** The first phase is the strategic orientation. Based on a roadmap or search fields, where strategic corporate goals and future trends and needs flow in, the topics are defined as guidelines. It is defined where - specifically in which market, product or technology fields - innovation takes place and which ideas are to be searched for.
2. **Inspiration:** Based on the strategy, brainstorming starts, which is an iterative process. There are a variety of sources for ideas and methods for generating ideas: employees, customers, LEAD users, suppliers, application of creativity and innovation methods, patents, new technologies, start-ups, crowdsourcing, idea competitions, pitching contests, hackathons, *etc.* (Lead user is a term developed by American economist Eric von Hippel. *LEAD users* are "advanced users" or users, inventors and pioneers who deal with problems in a specific area for which there is no solution on the market).
3. **Ideation:** Not all ideas can be implemented, but it is necessary to identify those who have the greatest leverage. In this phase, the evaluation criteria and methods are defined and applied. The aim is to select the most promising ideas for the innovation portfolio from a wide range of ideas.
4. **Proof of concept:** The most important contents of this phase are the knowledge of customer requirements, opportunities and market potentials as well as the evaluation of feasibility and risks. Here, all risks and opportunities as well as customers and the environment are analysed. All stumbling blocks should also be detected in this phase so that they do not become fatal in later phases.

5. **Pilot (Prototype testing):** Solutions are sought, evaluated and prototypes are developed from the best ones and tested on the basis of criteria for the fulfilment of requirements and feasibility. The test knowledge is then used to fine-tune the development of further prototypes until the final solution is ready.
6. **Roll out and production:** Once the solution has been developed, it is ready for implementation. The production process must be developed, plans must be drawn up, resources mobilized and service processes must be established.
7. **Feedback and Launch:** Ultimately, the new product or service must be marketed so that the customer can buy or use it. This includes the development of sales channels, marketing activities, advertising, *etc.* In addition to marketing, feedback from the market and customers must also be collected during this phase, which in turn flows into the further development and optimization of the product and also into production and marketing. This phase then goes into product lifecycle management.

2.4. Forms of Innovation

As indicated, innovation is a dynamic concept with varied forms evolved over time depending on the challenges and need. The common forms of innovation are briefly explained:

2.4.1. Social Innovation

Social innovations are new social practices that aim to meet social needs in a better way than the existing solutions. It refers to the process of developing and implementing new, effective solutions to solve social or environmental issues. Social innovation adds an extra dimension to innovation, sustaining economic and social growth. Social innovation refers to "the reconfiguring of social practices, in response to societal challenges, which seeks to enhance outcomes on societal well-being and necessarily includes the engagement of civil society actors".

Mumford (2002) stated that social innovation entails complex processes introducing new products, processes or programs that profoundly change the basic routines, resources, and decision-making processes, or beliefs of the system in which the innovation occurs.

Westley and Antadze (2010) defined social innovation "as a complex process of introducing new products, processes or programmes that profoundly change the basic routines, resources and authority flows, or beliefs of the social system in which the innovation occurs. Such successful social innovations have durability and broad impact".

Social innovation can be considered the process of applying novel market solutions to global social and environmental problems by creating or improving products, services, processes, business models and markets to more effectively and efficiently respond to unmet societal needs.

Different types of definitions of social innovation (SI) have generated two main schools of thinking in SI (Nicholls *et al.*, 2015). The first school focuses on new social processes. This relates to exploring changes in social relations, and emphasising changing power balances towards economic equity in society. The second school emphasises understanding of new social outputs and outcomes, and is primarily oriented toward dealing with market failures in the provision of public social goods.

In line with the second school, the OECD, 2010 states that: "Social innovation is distinct from economic innovation because it is not about introducing new types of production or exploiting new markets in itself but is about satisfying new needs not provided by the market or creating new, more satisfactory ways of insertion in terms of giving people a place and a role in production".

Types of Social Innovations

Dutch philosopher Herman Dooyeweerd designed a systemic framework identifying ten types of social innovation: socio-juridical, socio-cultural, socio-political, socio-ideological, socio-ethical, socio-economic, socio-organisational, socio-technical, socio-ecological and socio-analytical. Wigboldus (2016) provides an explanation of the ten types as shown in Box 2.1.

Concept of Social Innovation

There are three main interpretations of the social innovation concept, referring to:

- ✰ The social mechanisms of innovations,
- ✰ The social responsibility of innovations, and
- ✰ The innovation of society

1. Innovation is a social process between different actors. This is linked to the concept of social innovation. The concept of social innovation originates in critiques of traditional innovation theory. By calling for social innovation, new theories point at the need to take the social mechanisms of innovation into account.

   Introduction of new products and new technologies depend on its fit into a specific social context with a specific organisation of social relations and specific norms and values and accepted behaviour patterns. Recent theories about innovation use the concept of socio-technical innovation to explicate the inseparability of the social and technical in processes of innovation (Smith *et al.*, 2016).

Box 2.1

1. Socio-juridical innovation- innovation of legal frameworks, laws, *etc. e.g.,* Citizen jury introduced in Dutch courts
2. Socio-cultural innovation- innovation of non-formal institutions. *e.g.,* World Economic Forum
3. Socio-political innovation- innovation of governance, policies, *etc. e.g.,* Social Audit of Government programmes in India
4. Socio-ideological innovation- innovation of ideological frameworks, mindsets, paradigms, *etc. e.g.,* Self Help Group (SHG) movement in India
5. Socio-ethical innovation- innovation of ethical/normative frameworks. *e.g.,* Corporate Social Responsibility (CSR)
6. Socio-economic innovation- innovation of economic models, business models, *etc. e.g.,* Value Chain Analysis (VCA) of crops
7 Socio-organisational innovation- innovation of organisational/ institutional arrangements *e.g.,* Farmer Producer Organisations (FPOs)
8. Socio-technical innovation- innovation of human technology and interaction *e.g.,* Social media
9. Socio-ecological innovation- innovation of human environment and interaction *e.g.,* System of Rice Intensification (SRI)
10. Socio-analytical innovation- innovation of analytical and sense making frameworks *e.g.,* People Science movements

Source: Wigboldus, S A (2016). Ten types of social innovation – a brief discussion paper. *https://edepot.wur.nl* ›.

The construction and introduction of new technologies always involves changes in the interaction of things (artefacts), actors and ways of doing (institutions) and affects and is affected by how society is organised and functions. This is most evident in the case of 'system innovations' that go beyond the introduction of a new product or process but change the context, manner and meaning of how something is done, and lead to fundamental changes in many areas of society.

2. Innovations must take into account a social responsibility. Innovations should not only focus on the profit aspect but also on the planet and profit aspects of sustainability (the social responsibility of innovation). As innovation is also disruptive, this can be a challenging demand.

   In classic economic thinking innovation is considered important because of its ability to increase profit and encourage economic development. Even today, innovation is often associated with industries developing new products and new technologies driven for profit maximization. Technological innovation is also increasingly met by scepticism and concern about potential risks for human safety and the environment. The debate about genetic modification is a well-known example for these concerns. All this may be summarised under a call for social or socially responsible innovation: innovations that are ethically approved, socially acceptable and relevant for society.

3. Not only commercial activities, but social and public activities also need innovation. In the context of rural development, social innovation refers to the social objectives of innovation-those changes in the social fabric of rural societies, perceived as necessary and desirable in order to strengthening rural societies and addressing the sustainability challenge (social inclusion/equity: the innovation of society as well as the social responsibility of innovations).

Social innovation is referred to when indicating the need for society to change as a prerequisite for solving pressing problems such as discrimination, poverty or pollution. The focus is on changes in social relations, people's behaviour, and norms and values. Social innovation is often interchanged and used with concepts such as social empowerment and inclusion, social capital and cohesion. However, Stanford Centre for Social Innovation deviates from such an interpretation.

Box 2.2

SIMRA (Social Innovation in Marginalised Rural Areas) is a project funded by the European Union's Horizon 2020 programme. (Horizon 2020 was the EU's research and innovation funding programme from 2014-2020. The programme has been succeeded by Horizon Europe).

It aimed to advance understanding of social innovation and innovative governance in agriculture, forestry and rural development, and how it can be boosted, in marginalised rural areas across Europe and around the Mediterranean, including non-EU countries.

Social Innovation Actions constitute a new approach for hands-on practice and implementation of research. Initiating social innovation actions is a complex social process, which requires conceptualisation, testing and implementation of actions, both user-centred and context-specific. Social Innovation Actions enable that relevant actors in marginal rural areas test and exploit their potential for social innovation, with the aim of realizing impacts in the territory and the market, in businesses, investments opportunities and in building capacity within local governments.

Source: http://www.simra-h2020.eu

By stressing the need to include and give voice to socially deprived groups, the political element of innovation comes to focus. In any innovation processes, it is important to keep a close eye on who are considered to be included in the innovation processes, and who are eventually to gain or lose from the changes brought about. Social innovation is also strongly related to the innovation of politics and governance.

Characteristics of Social Innovation

The process of social innovation is characterized by:

- ✰ Changes in social relations that increase the level of satisfaction and participation of social groups, especially private groups.
- ✰ Strengthening of social movements and initiatives and empowerment of individuals and communities.
- ✰ Satisfaction of human needs either not met or perceived as important by the market or the state.
- ✰ Plurality of ideas contributing to a more complete representation of the problem, its causes and possible solutions.
- ✰ Co-operation between various actors through strategic alliances, partnerships, networks and multidisciplinary teams
- ✰ Promotion of learning and knowledge creation
- ✰ Adoption of holistic approaches to social problems
- ✰ Development of solutions aimed at local communities
- ✰ Establishment of a new mode of governance for learning
- ✰ Generation of social and humanitarian values by improving well-being as well as social justice, inclusion, cohesion and integration
- ✰ Changes in attitudes, core values, strategies, policies, structures, organisational processes, public services, and working conditions.

Box 2.3

The Centre for Social Innovation under the Stanford Graduate School of Business (SGSB) educates insightful leaders for social and environmental change. The social innovation curriculum is part of a holistic social-innovation learning model that builds off of the management core curriculum and leverages academic frameworks, students' and professionals' experiences, and hands-on discovery. The Centre support the faculty in developing courses grounded in the fundamentals of social innovation: impact measurement, human-centred design, systems thinking, and cross-sector approaches. Stanford GSB faculty contribute new thinking on ways to achieve social impact, and the Centre provide research services and support to faculty conducting research on a variety of topics related to social innovation.

Source: http://gsb.stanford.edu/experience/about/centers_institutes_csi

Social Innovation Models

Some of the models of SI are presented:

1. Microcredit as a Social Innovation

To be considered an innovation, a process or outcome must meet two criteria. The first is novelty. Although innovations need not necessarily be original, they must be new to the user, context, or application. The second criterion is improvement. The process or outcome must be either more effective or more efficient than pre-existing alternatives. To this, one more criterion is added- sustainability. By sustainable, we mean solutions that are environmentally as well as organizationally sustainable, those that can continue to work over a long period of time.

The difficulties of lending to the poor are- lack of complementary inputs for the poor to productively use capital, risks and market imperfections such as information asymmetry, as well as transaction costs. The transaction costs become high since the loan size is low. The risks are high, owing to asymmetric-information-related problems of adverse selection. Before giving the loan, the lender is not able to distinguish between safe and risky borrowers due to lack of credit history.

In any case, it was believed that the poor were unable to use the financial capital productively since they did not have the social and human capital, required as complementary inputs. For any social innovation to work, it had to offset the existing problems.

Microcredit is an example of social innovation because it is an initiative taken by actors in order to provide new answers to social problems (social and financial exclusion, poverty) and a device capable of providing services to enable construction of a society with high levels of quality of life. It is considered as an innovative system of lending aimed at financial inclusion of the poor.

Microcredit is a social innovation, using socially available information, peer lending, social collateral, and public payments to overcome problems of information asymmetry that can cause high-loan-default rates. When microfinance benefits women, it contributes to empowerment, well-being and development.

2. Fairtrade as a Social Innovation

Fairtrade initiative innovated as a social movement by offering an international trade model to make difference in small producers' lives and livelihoods from developing countries. According to Fairtrade Foundation, it challenges the conventional model of trade and offers a progressive alternative for a sustainable future.

What is novel about fair trade is that it works at so many links in the value chain, from farmers to salespeople to consumers. The model not only is novel, but it also creates tremendous social and environmental value by deploying a host of safeguards, including sustainable agricultural techniques, international certification and labelling, child labour prevention, and fair prices. Fair trade also generates significant economic value.

While pioneer Fair Trade Social Enterprises in the early years were largely non-profit organizations relying on voluntary work, they have now become increasingly diversified in terms of legal forms, governance models and organizational practices. These diversified models reflect the hybrid nature of FT, through different ways of combining a commercial activity (trading of FT products), a social mission (support to producers), and a strong political message on decent livelihoods (often expressed through education and advocacy).

3. Under The Mango Tree (UTMT) as a Hybrid Social Innovation Model

UTMT Society, founded in 2009, promotes beekeeping with indigenous bees like *Apis cerana indica* and *Trigona* to increase agricultural productivity and enhance biodiversity, incomes and livelihoods of marginal farmers in India. UTMT Society works in tribal areas in Gujarat, Maharashtra and Madhya Pradesh in India. UTMT aims to promote community-based beekeeping as a specific solution to declining agricultural productivity and improvement of livelihoods of marginal farmers across India.

Under The Mango Tree works at both ends of the honey supply chain. Through its for-profit, UTMT Pvt. Ltd., it sources high-quality organic honey from farmer cooperatives, before packaging it for sale direct to customers. Through the for-profit, Under The Mango Tree Naturals and Organics Private Limited, the following activities are done.

- ☆ Establish a fair-trade market for locally produced honey that is natural, organic and sustainable.
- ☆ Source high-quality organic honey from farmer cooperatives.
- ☆ Package the honey for direct sale to customers.

Through the non-profit, UTMT Society, activities undertaken are:

- ☆ Train and equip smallholding farmers to use bees to increase their incomes by as much as threefold.
- ☆ Implement a buy back arrangement for honey produced, which increases farmers' income by another 40 per cent from the sale of their own honey.
- ☆ Realize the goals of Bees for Poverty Reduction and Urban Beekeeping Programmes. Engage in research and policy advocacy to advance sustainable community-based beekeeping with the indigenous bee, the *Apis cerana indica*.

Social Capital and Innovation

Social capital refers to features of social organization, such as networks, norms, and trust, that facilitate coordination and cooperation for mutual benefit. The notion of social capital extends our understanding of "cooperation" or "collaboration" in two significant ways. First, linking cooperation to the economic concept "capital" signals the investment or growth potential of a group's ability to work jointly. Second, the concept identifies the structure created from collaborative effort as capital. Well-functioning partnerships, consortia, and networks are in and of themselves "a form of social capital." Capital is located both in the sharable resources held by individual institutions in a network and in the overall structure (relationship) among the institutions in a network.

Social capital is a necessary, although not sufficient, enabler of effective public-private partnerships, for devolution of some science and technology responsibilities to the states, and for a new, more collaborative style of government policy. The fundamental significance of social capital for innovation, and thus for science and technology policy, has yet to be assimilated by policy-makers and captured in the design of policy tools.

As the pace of technological change has intensified, as economic resources have become more scarce, and as information technologies have made linkages among geographically dispersed actors commonplace, the predominant form of economic organization has been changing. Large, centralized bureaucracies emphasizing division of labour and functional specialization have given way to smaller, leaner organizations in which team-based structures cross functional lines, disrupt traditional hierarchical chains of command, and focus on core functions while contracting with outside firms for other tasks.

The biotechnology industry provides a cutting-edge example of new forms of industry operation that demand an intensive level of external collaboration. The traditional view of technological development and industry structure holds that technologies follow a typical life cycle whose character shapes the form of an industry. In industries as diverse as high technology, heavy manufacturing, and utilities, the trajectory of technological development and change and its effect on industry structure have been viewed by scholars as roughly similar.

One of the best-known examples of a high-performing industry network is the computer industry of Silicon Valley, California. The professional culture is highly collaborative. Non-proprietary professional and technical information is regularly shared among employees and companies.

Three different, but highly interrelated, structures- firm, industry, and institution- work together in Silicon Valley and explain the high levels of efficiency, adaptability, and innovation within this setting. The system exhibits congruence, or alignment, among the internal firm structure, the organization of the network, and the wider institutional structure in the region that supports the industry.

Social Innovations in Agriculture

Social innovation is popular in the context of agriculture and rural development but the use and importance attached to it differ according to the domain and scope of innovation referred to.

In addition, it has a considerable political or normative weight. First of all, social innovation is most frequently used in the context of rural development, where the need for social changes is most evident. When rural development is concerned, the 'social' is presented as a core element of innovation, and in the sense of engaging society in developing new solutions.

When it comes to strictly agricultural development in relation to production efficiency, social innovation is generally considered less significant. Here a technology-oriented definition of innovation predominates. This has also to do with the different scope of innovations. Agricultural development, as such, often deals with innovations in the sense of new products or new processes whereas rural development regards the innovation of socio-economic systems.

In the past, social mechanisms were considered as important when reaching the phase of diffusing innovations, when experts transferred new knowledge, products and/or technologies to users and convinced them to accept and use them. Traditional Agricultural Knowledge Systems are based on this approach.

The call for what might be framed as responsible agri-rural innovation is received in various ways, reflecting different approaches to innovation. At the one hand we see attempts to meet social concerns by way of new technological designs, that reduce the negative effects. This is often achieved through more efficiency and reduction in either energy demand or polluting emissions (*e..g.* precision agriculture).

In the field of agriculture and food and nutrition security, different forms and cases of SI can be highlighted, as below:

- Processes of resilience and adaptation in production of food;
- Inclusive participation and new roles for stakeholders;
- Community-led organisations and bottom-up initiatives;
- Different interpretations and usages of technologies;
- New kinds of agro-food partnerships;
- Citizen science initiatives.

Successful Cases of Social Innovations in Agriculture

1. System of Rice Intensification (SRI)

In India, the SRI is a good example for social innovation in agriculture. In this case, a combination of factors such as the historical influence of Gandhi's independence teachings to search for alternative narratives, availability of community spaces for exchanging indigenous knowledge, a deep crisis in farmer agency (high productivity rates but low profits leading to a high percentage of farmer suicides), and a heavy emphasis on green revolution technologies led to a counter-response among the farmers. The system did not involve any genetic modification.

The idea behind SRI involves managing rice plants, soil, water, and nutrients with reduced use of material inputs while creating productive and resilient varieties in a collaborative manner. These approaches were tested in collaboration between researchers, civil society organisations, and farmers. While initially meeting with heavy resistance from agricultural research and extension institutes, eventually SRI became an accepted body of knowledge on crop intensification and helped to foster diversity in thinking and renewed valuation of local experimentation and community engagement.

2. Farmer Field Schools (FFS)

Piloted by the United Nations Food and Agriculture Organisation since the early1990s, Farmer Field Schools were developed as an approach countering top-down Green Revolution extension methods in places where complex or contradictory problems challenged farmers. The premise of putting farmer peer-to-peer learning and group experimentation ahead of technical knowledge constituted a new approach to farmer capacity development. This approach allowed farmers to investigate, test, and decide for themselves what production methods worked for their environment.

FFS is a group-based adult learning approach that teaches farmers how to experiment and solve problems independently, also called "schools without walls". In FFS, groups of farmers meet regularly with a facilitator, observe, talk, ask questions, and learn together. As an extension approach, FFS differs from the traditional, top-down "transfer of technology" method. Farmers interact with researchers for help only when they cannot solve a problem by themselves.

3. Family Farming as Social Innovation

The multi-functionality of agriculture is a concept based on the increasingly segmented demand of goods and services that comes from society. The family farm, then, becomes a novelty when it embraces the paradigm of multi-functionality and starts a production process oriented to achieve scope, not scale economies (Milone, 2009). Milone describes how farmers develop novelties by using endogenous resources that are strongly linked to the territory and

Box 2.4

The UN Decade of Family Farming (2019-2028) is a joint effort by the Food and Agricultural Organization (FAO) of the UN and the International Fund for Agricultural Development (IFAD), both of whom are members of the Global Donor Platform for Rural Development.

Family farmers have huge potential to become key agents of development strategies, such as helping to ensure that food systems contribute to healthier diets and ensuring the environmental, social and economic sustainability of our food systems. Supporting family farmers also means managing land rights and developing enabling policies and investment plans.

The seven pillars of the UN Decade of Family Farming are:

1. Develop an enabling policy environment to strengthen family farming
2. Support youth and ensure the generational sustainability of family farming
3. Promote gender equity in family farming and the leadership role of rural women
4. Strengthen family farmers' organisations and their capacities to generate knowledge, represent farmers' concerns and provide inclusive services in rural areas
5. Improve socio-economic inclusion, resilience and well-being of family farmers, rural households and communities
6. Promote sustainability of family farming for climate-resilient food systems
7. Strengthen the multidimensionality of family farming to promote social innovations contributing to territorial development and food systems that safeguard biodiversity, the environment and culture

Source: www.fao.org>family_farming_decade>home

embedded in local networks. The territory also functions as a protected space, a niche in which the novel products are sold in the local market, from where the products may eventually enter geographically distinct markets. To attain this purpose, rather than increasing the level of production, and the size of farms, priority is given to enhance a number of factors and conditions that distinguish a multi-functional and multi-product enterprise, which coincides with a renewed farm model. These factors are:

- ☆ Diversification of business activities and production;
- ☆ Flexibility of family labour (multiple jobs);

- Acquisition of knowledge and skills of farmers;
- Prevalence of artisan component in the production process (local resources);
- Creation of a reference market;
- Creation of a network through which to intercept external markets even outside the regional limits.

2.4.2. Grassroots Innovations

Grassroots innovation is a diverse set of activities in which networks of neighbours, community groups, and activists work with people to generate bottom-up solutions for sustainable developments; novel solutions that respond to the local situation and the interests and values of the communities involved.

Grassroots innovations are community-led solutions for sustainability. They can offer promising new ideas and practices, but often struggle to scale up and spread beyond small niches. There are many grassroots innovations found successful globally. Some well-known cases are presented:

1. The Movement for Socially Useful Production (UK: 1976–1986)

In 1976, facing the cuts of thousands of jobs, workers at the British company Lucas Aerospace published an Alternative Plan for the future of their corporation. The Lucas Plan is one of the key artefacts of the movement for socially useful production, which flourished for a brief period in the UK in the 1970s and 1980s.

The Plan assumed a symbolic role within a wider critique of mainstream policy towards technology and economic development. The proposals in the Plan became symbolic for a movement of activists committed to innovation for purposes of social use over private profit. An important highlight of the Plan, human-centred technologies that enhanced skills rather than displaced labour, caught the attention of some activists and broader arguments for socially shaping technology for community benefit attained prominence with the emerging radical science movement.

However, the movement and activities were swept aside by the rise of Thatcherism and the installation of neo-liberal ideology. The notions of social purpose were overwhelmed by the ideology of free markets. But the real legacy of the movement has been its contribution to the wake up call for more democratic institutions for the social shaping of technologies and innovation. The outcome of the movement were:

- The movement found its first expression in the workplace, where technological change, particularly computer integrated manufacturing (CIM), was seen to be deskilling and displacing workers.

- ☆ Conventional trade unionists identified their task as first resisting, and then negotiating a share of the productivity gains arising from this trajectory of development.
- ☆ Socially useful production meant bringing local community groups and activist networks into the production process.

2. The Appropriate Technology Movement (South America: 1970s and 1980s)

Appropriate technology is defined as any object, process, idea, or practice that enhances human fulfilment through satisfaction of human needs. It is an ideological movement originally articulated as intermediate technology by the noted economist E. F. Schumacher in his influential work, "Small is Beautiful". Schumacher (1973) set a direction for a new technological order when he remarked, 'What we need is a technology with a human face'. It is generally recognized as encompassing technological choice and application that is small-scale, decentralized, labour-absorbing, energy-efficient, environmentally sound and locally controlled.

A technology is deemed to be appropriate when it is compatible with local, cultural, and economic conditions and utilizes locally available materials and energy resources, with tools and processes maintained and operationally controlled by the local population. Technology is considered thus "appropriate" to the extent that it is consistent with the cultural, social, economic, and political institutions of the society in which it is used. It represents the social and cultural diversions of innovation.

Appropriate technology movements can be found in both developing and developed countries. In developed countries, the appropriate technology movement grew out of the energy crisis of the 1970s and focused mainly on environmental and sustainability issues.

Appropriate technology has been used to address issues in a wide range of fields. A well-known example of appropriate technology application in agriculture is the use of pedal-powered water pumps. The India Mark II hand pump designed to lift water from a depth of fifty metres or less, is world's most widely used water handpump. The pump was designed in the 1970s to serve village water needs and has been the central point of successful community based water supply system in India as an appropriate technology which covers most rural areas in India.

Today appropriate technology is often developed using open source principles, which have led to open-source appropriate technology (OSAT) and thus many of the plans of the technology can be freely found on the internet. OSAT has been proposed as a new model of enabling innovation for sustainable development.

3. The People's Science Movement (India: 1960s to present)

All India People's Science Network (AIPSN) is one of the People's Science Movements (PSMs) formed in 1988 with 26 member organisations, that include Chipko Movement, Eklavya, Friends of Rural Society, Kerala Sastra Sahitya Parishad (KSSP), Kishore Bharati, Movement in India for Nuclear Disarmament (MIND). Today it has more than 40 members spread across the country.

AIPSN address the question of the role of science and technology in furthering the goals of social and redistributive justice for the marginalised people and communities. Several initiatives such as People's Technology Initiative and the People's Health Movement (Jan Swasthya Abhiyan), are organised in broader coalition with other agencies.

In the past few decades, PSMs have influenced "decision making in science and technology on the environment, big dams, science education, energy and GMOs" and have become an "important actor effectively intervening into national science and technology systems and decision making in science" (Krishna, 2014).

The enterprise of science, a driver of productive forces, is supported by capitalism. In this framework of functional ideology, the ecological threats and deprivations are explained as a natural consequence of 'profit-first' economic system. The destructive effect of the application of modern science and its inability to address specific questions facing the marginalised communities cannot be considered the result of the intrinsic malady of modern science and technology, but is a consequence of elite hegemony.

PSMs articulate their goal "to build a secular and democratic public sphere that will provide the opportunities for free scientific debate and resolution of people's problems on a fully democratic and scientific basis, untainted by the hegemony of corporate capital and sectarian community interests".

Using the framework of 'science for social revolution' (social innovation paradigm), PSMs are constructing an India-specific tradition of 'democratising science activism' on the fronts of education, development, and environment.

4. Hackerspaces, Fablabs and Makerspaces (International: 2000s to present)

Though Hackerspaces, makerspaces and fablabs were developed independently, each have more or less converged towards a similar structure and use. Each can be characterized as a community workshop. They can be considered community oriented physical places, where people share their interest in tinkering with technology, meet and work on their projects, and learn from each other.

Makerspaces, hackerspaces and fablabs arise from distinct origins and that is why they possess different names. The term 'makerspace' is attached to community workshops where members share tools and was formulated in contrast to hackerspaces, considered more focused on computers and electronics. Fablabs originated directly from the MIT's Centre for Bits and Atoms. Fablabs were established as a model for digital fabrication workshops. The Fab Foundation website denotes four qualities of fablabs that must be present for an organisation to use the name- public access, support and subscription to the fab lab charter, a common set of tools and participation in the network of fablabs.

5. The Social Technology Network (Brazil: 2000s to present)

In the 1990s, the NGOs reached a new level pushed by some milestones like the United Nations Environment Summit that took place in Rio de Janeiro (1992), and the emergence of the Third Sector. The agenda of the NGOs also shifted to struggle for new rights such as the access to public information and the modernization of the NGOs.

In this scenario, part of the Brazilian NGOs started to work with new small-scaled technological models based on three assumptions: (1) Technology is not neutral since who produces has its own values and beliefs; (2) The current technological model is environmentally unsustainable; (3) Technology can be harmful causing problems such as air pollution, unemployment and alienation of the workers. The inspiration to this movement was given by Schumacher (1973) with the concepts such as "intermediate technology", "small scale technology" or "appropriate technology".

With the arrival of the new centre-left government of the Workers Party in Brazil, the Social Technology Network (STN) was formed in 2004 aimed at fostering a process of social inclusion, public participation and income generation by drawing from existing capabilities in Science and Technology.

Social Technology Network as a network articulation gathered more than 900 organizations (as NGO´s, Associations, and Cooperatives) that proposed to organize, articulate and integrate themselves with the purpose of promoting sustainable development through the diffusion and scale reapplication of social technologies.

Social technology refers to any technology that facilitates social interactions and is enabled by a communications capability, such as the Internet or a mobile device. Examples are social software (*e.g.*, wikis, blogs, social networks) and communication capabilities (*e.g.*, Web conferencing) that are targeted at and enable social interactions.

6. The Honey Bee Network (India: 1990s to present)

Established in 1986-87, The Honey Bee Network (HBN) is a volunteer based network in India that seeks innovative ideas and Traditional Knowledge (TK) produced at the grassroots level by individuals and communities and disseminates them to the wider ecosystem. The network was founded by Prof. Anil Gupta in 1988 as a social justice initiative to fix the failures of top-down development initiatives led by the government. HBN's primary purpose is to support grassroots innovators who are rich in knowledge, but not in resources. It is one of the pioneering organizations in India that focused principally on the poor.

HBN is an amalgamation of like-minded people, such as innovators, farmers, scholars, academicians, policy makers, entrepreneurs, institutions and civil society. The Network signifies a philosophy of discourse that stands on the three pillars: the discourse that is authentic, accountable and fair.

The Honey Bee Network's central objective is ethical knowledge extraction. It maintains a grassroots knowledge database that collects and disseminates expertise from a wide range of individuals while observing ethical practices of credit, compensation, and accessibility. It strives to build a network that allows the benefits of its centralized database to be shared with knowledge contributors that may normally go ignored under the traditional practices of knowledge extraction.

SRISTI (Society for Research and Initiatives for Sustainable Technologies and Institutions) is a developmental voluntary organization aiming to strengthen the creativity of grassroots communities, and individual innovators. It supports eco-friendly solutions to local problems being scouted, spawned and spread by the Honey Bee Network.

There are five pillars of Honey Bee Network which SRISTI is committed to reinforce:

1. Educational innovations by school and college teachers, students and other stakeholders
2. Institutional innovations at community and other levels in managing resources
3. Improving access of knowledge-rich, economically poor people to trigger self-reliant development process
4. Cultural creativity so that curiosity, collaboration, and compassion grow through art, literature and crafts *etc.*
5. Technological innovations and traditional knowledge dealing with human, animal, plant and ecosystem health, and policy reforms to generate frugal innovations for sustainable development at all levels, with specific reference to youth, children, women and elderly.

2.4.3. Corporate Innovation

Corporate innovation is the process of an enterprise experimenting with modern practices, technology, strategies, and opportunities to make its existing business model more competitive in the current marketplace.

Corporate innovation refers to out-of-the-box thinking as a strategy within the business. It can refer to product innovation, division innovation or business

Box 2.5

The traditional attitude toward creativity in the American business world is to "think outside the box", meaning to brainstorm without restraint in search of coming up with a breakthrough idea, often in moments of crisis. Though it works sometimes, it is a problem-specific solution that does nothing to engender creative thinking more generally. 'Inside the Box' demonstrates Systematic Inventive Thinking (SIT), which systemizes creativity as part of the corporate culture. This counterintuitive and powerfully effective approach to creativity requires thinking inside the box, working in one's familiar world to create new ideas independent of specific problems. SIT's techniques and principles have instilled creative thinking into such companies as Procter and Gamble, Johnson and Johnson, and other industry leaders. Inside the Box shows how corporates have successfully used SIT in business settings as diverse as medicine, technology, new product development, and food packaging.

Source: https://www.viima.com>blog>outside-the-box-innovation

model innovation. Division innovation is a tool of systematic inventive thinking (SIT) method. The method is described in the book "Inside the Box" by Boyd and Goldensberg.

Currently, most digital innovations in agriculture are led by 'Big Ag' companies, smaller innovative agricultural technology (ag-tech) companies, and top agricultural universities. However, corporate innovation in digital agriculture technologies is mostly associated with a few global-scale companies that offer durable (farm equipment) and consumable (seed, chemicals, *etc.*) goods and services. These industries have in recent years consolidated to the point where most major farm purchases are controlled by a handful of companies in a highly competitive global market. These corporate leaders are primarily headquartered in Northern America and Western Europe and increasingly differentiate themselves in the marketplace by their ability to innovate with digital technologies. Yet smaller companies, typically based in the same countries, also offer innovative technology solutions. A good example for corporate innovation in agriculture is Digital Green.

2.4. 4. *Jugaad* and Frugal Innovation

Jugaad is nothing but a creative problem-solving process, by utilizing the available resources. It solves the problems in an innovative way and can be easily replicated by any common individual. It is a commonly used word in India and a lot of people practice it in their day to day life to solve small issues or to survive from any crisis. Many of these practices can be frequently seen in automobiles and agriculture sector.

Box 2.6

Digital Green is a global development organization that empowers smallholder farmers to lift themselves out of poverty by harnessing the collective power of technology and grassroots-level partnerships. Digital Green strongly believe that technology can accelerate efforts to end poverty. It provides training and technical support to its partners and develops technology-based solutions to empower rural farmer communities. It is supported by the Bill and Melinda Gates Foundation, USDA, Google, Oracle and Cisco.

Digital Green has defined standard operating procedures and has a variety of technology tools that are open source and can be customized. FarmStack is powered by Digital Green and made possible through investments by the Bill and Melinda Gates Foundation and Walmart Foundation.

Kisan Diary is an Android app developed by Digital Green for farmers to capture, analyse and control their financial data and manage their farming and household expenses better. This first-of-its-kind personal utility app made specifically for farmers provides a free, safe and secure service in four languages. The app provides crop and time-based profit/loss analysis.

Loop is an app-based logistics service for farmers which explores potential models for farm to market linkages that can be made sustainable, with farmers willing to pay for the increased convenience and better prices. It is currently being spun off as an independent social enterprise.

Source: https://www.digitalgreen.org

Some of the characteristics of jugaad are: Quick solutions; Do more with less; Find replacement of materials, functions, and forms; Make it simple; Constraints of resources; No need for tech to solve all the issues; Maximum usage of available resources; Focus on the core functionality; Use the survival instincts and Think and act flexibly.

Jugaad is not the final solution, it is a temporary solution for the immediate needs, which may not be lasting. Often, jugaad might serve a quick relief. But it doesn't solve the problem for a longer period and the quality of the solution also goes down over time.

To take the Jugaad solutions forward, the critical aspects are: sustainability, large scale manufacturing, and maintaining quality with reliability. Mitti Cool, the low cost biodegradable refrigerator made out of clay is a good example for juggad.

Frugal innovation is a practice of rethinking the problem and existing solutions to come up with a sustainable and reasonably low-cost solution. It has more functions than the other existing solutions and can be easily manufactured at scale. It is about leveraging the limitation in resources like financial, materials, or institutional and by using several methods, turning these constraints into the advantages.

Successful frugal innovations are characterized by:

- ☆ **Low-cost products or services-** It extends to services as well, not just products. It is about making better things not just cheaper things. It reduces the cost drastically to about one-third of the currently existing solutions.
- ☆ **Outperforms the alternatives:** It gives more in terms of performance and other use cases. It is about remodeling, not de-featuring. Higher performance does not always mean higher specifications.
- ☆ **Available at large scale:** It easy to manufacture at a larger scale and thus, it reduces the cost as well.
- ☆ **Quality and assurance:** It doesn't compromise with the quality as it reduces the cost. Low cost does not mean low-tech.

Very often frugal innovation products and services are derived from a social mission of serving the bottom of the pyramid (BoP). Examples like Jaipur foot, Arvind eye hospital, Narayana Hridayalay, Tata nano had the social mission at its very core. Jaipur foot and Arvind eye's cataract surgery are serving a major portion of the society through free treatment and free products.

The characteristics like low cost, at par performance and availability, cover the essential requirements of social mission. The cheap cell phones made by local companies, prosthetic legs fashioned from irrigation piping, *etc.* are good examples for frugal innovation.

The increasing interest for various alternative innovations has also led to an avalanche of concepts such, Bottom Of Pyramid Innovation (Prahalad, 2004), Cost Innovation, Frugal Innovation, Jugaad Innovation, Inclusive Innovation and Resource Constrained Innovation (Sharma and Iyer, 2012).

In definitive, the philosophy of frugal innovation is based on the transformation of scarcity into competitive advantage through frugality, flexibility, ingenuity and inclusiveness. Frugal innovation concerns multiple innovators such as individuals, local firms, public organizations, multinationals, *etc.* It very often involves the development of technological products or services. If supported by suitable policies and endorsed by actors from developing countries, this approach can contribute to tackle brain drain experienced by those countries.

2.4.5. Local Innovation

Local innovation is the process by which people develop new and better ways of doing things, using their own resources and on their own initiative.

Local innovation often occurs in the face of new challenges or opportunities. Local innovations are usually developed and are understood and owned by local people. These innovations may be developed by individuals or groups or even the entire communities.

The traditional practices cannot be regarded as local innovations. Traditional practices are handed over from generation to generation and do not reflect an effort to find new and better ways of doing things. Local innovation takes place, however, when farmers adapt or modify traditional practices.

Similarly, a local innovation differs from an invention. A local innovation is new in a specific locality. It may already be practised elsewhere, but it is new in local terms. A farmer who has developed an innovation has not simply copied it from someone or somewhere else; s/he has come up with it on her/his own initiative. An invention, on the other hand, is a technique/technology that is new in absolute terms, *i.e.* it had never been "discovered" or developed anywhere else before.

Participatory Innovation Development

Participatory Innovation Development (PID) is an expansion of the Participatory Technology Development (PTD) approach (ILEIA, 1991). The Information Centre for Low External Input and Sustainable Agriculture (ILEIA) provides information to organizations and individuals about agricultural development in the Tropics. It can provide information about practical methods and the scientific principles on which such methods are based, as well as about specific management strategies by which conversion to more sustainable, low input farming may be achieved. It also documents Low External Input and Sustainable Agriculture (LEISA) and People-centred Participatory Technology Development (PTD) experiences and assesses those experiences in workshops.

At the heart of PID is "farmer-led joint experimentation"- or, more broadly expressed, "farmer-led joint research" - in which farmers together with support agents investigate possible ways to improve the livelihoods of local people. The word "research" suggests that the activities can go beyond field-based trials to include investigation of other questions, such as jointly exploring a value chain or trying out a different way of managing communal resources.

PID is an approach to research, extension and above all development. Often, it is undertaken by farmers together with development agents, without the involvement of formal researchers.

The words "Innovation Development" in PID do not imply that PID is confined to working with local innovations. Any joint experimentation anchored in community-level assessment and decision-making can be PID. Looking for local innovations is often a good starting point for PID, especially because it helps change the attitudes and behaviour of all involved.

2.4.6. Convivial Innovation

Convivial innovation is characterized by the need to slow down ("Low tech"). Such an innovation is not seen as a driver of growth, does not exert pressure, and in fact does not incorporate the concept of obsolescence.

Convivial innovation is adaptable, accessible, and locally available, and it is seen as a means to satisfy community needs. This concept is clearly different from the "market" and "technology" bias in mainstream innovation theory. The technological system converts means into ends (Illich, 1993).

User-friendly innovation is a way to increase individual and collective well-being, through the provision of expertise, innovative or updated techniques such as permaculture, but also various creations, such as eco-responsible construction, biodynamic agriculture, solar energy and home-made clothing, *etc.* It enables the rediscovery and revaluation of ancient knowledge.

According to Illich, "a convivial society is one that gives individuals the possibility of exercising the most autonomous and creative action, using tools that are less controllable by others. Productivity is understood in terms of **having**, conviviality is understood in terms of **being**. The tool is convivial in the sense that anyone can use it, without difficulty, as often or as seldom as they wish, for purposes that they themselves determine."

One pragmatic alternative to the subjugation of skilled practice by dehumanizing technological devices is the adoption of a cultural philosophy that recognizes the social value of convivial practices. Conviviality is a way of looking at the world that transcends the economic pragmatism of the technological paradigm. Conviviality illuminates the interrelatedness and complexity of living inextricably within social and natural systems.

To adopt conviviality as part of one's way of taking up with the world encourages graceful participation in something larger and more important than the mere instrumentality of the technological paradigm. A practice is convivial, if 1) it is inherently enjoyable as a creative process and not merely a means to an end, 2) it encourages and rewards companionship and self-directed cooperative interaction between citizens, and 3) it employs tools that enhance the social and natural environment.

First, conviviality entails engagement in a practice that is not solely based upon rationality, specialization, and standardization. Convivial work enlarges and illuminates the range of personal competence and control within a local setting and gives a practice a uniqueness created by a mixture of individual skill and local sources of energy. A convivial practice fosters a high degree of independence because conviviality requires that tools and skills should be firmly within personal control. Tools must be accessible and comprehensible by all who wish to use them. Third party control of tools is inconsistent with the principles of conviviality because it leads to centralization and radical monopoly, both of which create dependency upon a body of managers and specialists that is philosophically incompatible with the goals of a convivial practice.

Second, conviviality applauds social relationships and dependencies that are built around mutual participation and exchange of skill and energy within an intimate social environment. People have the necessary independence to exercise skill while also maintaining a cooperative spirit and working to achieve a high level of personal satisfaction rather than impressive production statistics.

Third, conviviality breeds dignity and well-being within society, because practices are meant to enhance social and ecological relationships. The attractiveness of a convivial practice is closely linked to the use of tools which allow a person to be conscious of the way he/she utilizes personal wisdom, skill, and discipline or cooperates with another to perform a practice.

Conviviality is simply one way of taking up with the world. While the technological paradigm proposes to intervene in our lives and do things for us, conviviality argues for the preservation of practices which require personal accountability. Convivial practices are true to social and ecological processes and do not lend themselves to shortcuts, as the desire for shortcuts is the wellspring of the technological paradigm. Further, the technological paradigm erodes the physical and intellectual capabilities and renders us incompetent to think and act for ourselves. The social and ecological implications of the technological paradigm are frightening, which needs a close examination.

In 2017, Andrea Vetter extended Ivan Illich's tools of conviviality by designing the five- dimensional friendly tools "Matrix":

- The relationship with the environment, an integral part of the ecosystem;
- Accessibility facilitated by the sharing of tools;
- Adaptability to local context and scale;
- Organic-interaction and regenerative capacity;
- The appropriation of locally available materials and resources, such as bicycles, houses and sewing machines, *etc.*

2.5. Risks of Innovation

Innovation is a complex process with many different stages and inherent risks. It often requires an investment of time and money before any positive effects become visible. Though innovation is the lifeblood of any organisation, risk is one of its fundamental dangers. A careful balance of the two, and collecting as much knowledge as possible, will help better innovate and take the big, smart risks when they really matter.

There are several **types of innovation risks** faced in business. Risks can be:

- ☆ **Operational** – *e.g.* failing to meet the quality, cost or scheduling requirements
- ☆ **Commercial** - *e.g.* failing to attract enough customers, and
- ☆ **Financial** - *e.g.* investing in unsuccessful innovation projects

The types of risk that can impact the success of an innovation are market risk, credit risk, operational risk, strategic risk, liquidity risk, regulatory risk, reputational risk, political risk, ***etc.***

Time plays a big role in developing risks. The longer the innovation process, the higher the risk associated with it.

Another classification of risk is in terms of Product risk, Technology risk, Team risk and Market risk.

Product Risk: Will the product be better than what exists? Will it be more successful in helping customers reach their goals? Will it be more useful, more usable, more cheap, *etc.*?

Technology Risk: Will it work? Will the new technology be reliable enough, fast enough, good enough? Will the technology be financially and operationally sustainable?

Team Risk: Does the company have the right people? Can this team fund the company, hire well, validate the other risks, and execute?

Market Risk: Will anyone care? Will people use the product? Will we be able to reach customers? Will the market trust us? Is the market as large as we think it is? Is it growing?

Risk is also classified as:

1. Value risk (whether customers will buy it or users will choose to use it)
2. Usability risk (whether users can figure out how to use it)
3. Feasibility risk (whether our engineers can build what we need with the time, skills and technology we have)
4. Business viability risk (whether this solution also works for the various aspects of our business)

The ideal innovation process is the trifecta of desirability, feasibility and viability, an idea that originated from IDEO in the early 2000's (https://www.ideou.com/blogs/inspiration/how-to-prototype-a-new-business). If a business idea meets all three criteria, then it contains these essential characteristics:

- ☆ A desirable solution, one that the customer really needs
- ☆ A feasible solution, building on the strengths of the current operational capabilities
- ☆ A profitable solution, with a sustainable business model.

But if any one of the above is found missing, implementing the idea becomes riskier and costlier.

2.6. Scaling Up Innovation

Scaling of innovations is a process that leads to widespread use of an innovation. It is regarded the last step after the discovery, proof of concept and piloting of an innovation. In business it is often used as maximizing operational scale of the product.

Scaling up happens when a proven idea, be it a product or service, passes the proof of concept (PoC) stage and is deemed ready for use by many users across the organization. There should be a definite objective to produce business benefits, such as reduced costs, streamlined processes, increased productivity and a boosted market position. Scaling is a big step and a large investment for organizations. It requires budget and careful planning.

For organizations to succeed in an increasingly competitive digital economy, they must be capable of scaling-up an idea to a viable product or service. However, many are observed falling at the final stage due to a lack of management alignment, resources, skills and cultural understanding.

Organizations are not only trying to innovate, but are also creating what Gartner Research (2018) calls "pretty prototypes" that do not address real business challenges. According to Gartner Research, the organisations must focus on three elements to scale proven ideas successfully: ownership, behaviour and process. Each element contains a scale breaker (a factor that derails scaling) and a scale maker (a factor that accelerates scaling). "Scale breakers" tend to resist change, while "Scale makers" encourage change. Turning 'scale breakers' into 'scale makers' requires strong ownership of projects. Here, project managers must take responsibility for on-going development and not consider it as a simple dotted-line role. They must avoid ideas being diluted and be prepared to bring in expertise where required.

The product or service must identify a need within the business to have any chance of scaling up. To this end, the minimum viable product (MVP) concept taken from Lean Startup methodology can be effectively used. The methodology stresses the impact of learning on new product and service development. Lean Startup is an approach designed to shorten product development cycles and rapidly discover if a proposed business idea will work.

An organization must be convinced a PoC is viable as scaling is expensive. There are still further challenges in the scaling stage that may include security, compatibility or integration issues that need to be overcome.

REFERENCES

Bock B B (2012) Concepts of Social Innovation. https://research.wur.nl › publications › concepts-of-social-innovation

Brown. T. (2008) Design Thinking Thoughts: Definitions of Design Thinking retrieved from https://designthinking.ideo.com/?p=49

Bram Peters, Marion Herens and Jan Brouwers (2018), Capturing Social Innovations in Agricultural Transformation from the Field: Outcomes of a Write-Shop), Sustainability 2018, 10, 4727; doi:10.3390/su10124727

CTA (1991) ILEIA - networking for sustainable agriculture. Spore 33. CTA, Wageningen, The Netherlands

Elena G. Popkova, Bruno S. Sergi (2022) Smart Innovation in Agriculture, Springer, Singapore

Elizabeth Hoffecker (2021) Understanding inclusive innovation processes in agricultural systems: A middle-range conceptual model, World Development Vol 140, April 2021, 105382

FAO, 2015 Building Resilient Agricultural Systems through Farmer Field Schools; FAO: Rome, Italy, 2015.

Gartner Research (2018) How to scale innovation beyond petty prototypes, 23 May 2018

Giuliana Aparecida, Santini Pigatto and Gianluca Brunori (2021) Social innovation in high-quality agricultural systems: metrics for assessing processes and outcomes, Innovation The European Journal of Social Science Research 34 (1): 1 DOI:10.1080/13511610.2020.1867519

Illich, I (1973) Tools of Conviviality, Harper and Row, New York

Jackson, T (2011) Prosperity without Growth, Routledge, London

Kallis, G (2018) Degrowth, Agenda Publishing Limited, New Castle

Krishna, V. V. (2014). 'Changing Social Relations between Science and Society: Contemporary Challenges'. *Science, Technology and Society* 19 (2), pp. 133–159. https://doi.org/10.1177/0971721814529876.

Milone, P. (2009). Agriculture in transition: A neo-institutional analysis (1st ed., 256pp.). Assen: Van Gorgum.

Mumford, M D (2002) Managing Creative People: Strategies and Tactics for Innovation, Human Resource Management Review, 10, 313-351

Mumford, D Michael (2010) Social Innovation: Ten cases from Benjamin Franklin. https://doi.org/10.1207/515326934CRJ1402_11

Moses, Ma (2012) The Evolution of Innovation. http://psychologytoday.com/US/blog/the-tao-innovation/201204/the-evolution-innovation

Nicholis, A, Simon, J and Gabriel, M (2015) Introduction: Dimensions of Social Innovation, In Nicholis *et al.* (eds) New Frontiers in Social Innovation Research, Palgrave Macmillan. DOI:10.1057/9781137506801_1

OECD (2010) SMEs, Entrepreneurship and Innovation. https://www.oecd.org › cfe › smesentrepreneurshipandinnovation

Pyburn, R. and J. Woodhill (eds.) (2014) Dynamics of Rural Innovation - A primer for emerging professionals. LM Publishers, Arnhem

Prahalad, C K (2004) The Fortune at the Bottom of the Pyramid- Eradicating Poverty through Profits, Wharton School Publishing

Prasad, S.C. Innovating at the margins: The System of Rice Intensification in India and transformative social innovation. Ecol. Soc. 2016.

Roxana Bobulescu, Fritscheova A. (2021) Convivial innovation in sustainable communities: Four cases in France, Ecological Economics, Vol.181, March 2021, 106932 Smith, A (2014) Socially Useful Production, STEPS Working Paper 58, Brighton: STEPS Centre

Smith A, Mariano Fressoli, Dinesh Abrol, Elisa Arond and Adrian Ely (2016), Grassroots innovation Movements, Routledge, 2016

Sherwat E Ibrahim and Joseph Sarkis (2020) Technological innovations and degrowth opportunities from Urban Egypt: Initiating the Discourse, Policy Brief article. Front. Sustain. Cities, 16 Oct 2020,

https://doi.org/10.3389/frsc.2020.492866

SIMRA (2017) Collection of Examples of Social innovation, www.simra-h2020.eu

Spagnoli, Luisa and Mundula, Luigi (2017) The Family Farming: A traditional model to foster the Agriculture Innovation. https://iris.unica.it › retrieve › handle › BSGLg 69, 2017, 17-28

Schmelzer, M (2016) The Hegemony of Growth: The OECD and the Making of the Economic Growth Paradigm, Cambridge University Press, UK

Vetter, A (2017) The Matrix of convivial technology- assessing technologies for degrowth, J. Clean Production

Westley, Frances and Antadze, Nino (2010) Making a Difference: Strategies for scaling social innovation for greater impact, The Innovation Journal, Vol 15 (2).

Chapter 3

Innovation Ecosystem

An innovation ecosystem is a system for repeatedly delivering new, sustainable business models. It comprises of people, processes and systems that can support a concept from ideation to a scalable business. The term is used to describe the various players, stakeholders and community members critical for innovation.

An innovation ecosystem is not built on a list of people and places (universities, accelerators, angel investors, fab labs). It is built on the relationships between them, the exchange of information and resources that make a sustainable system. An innovation ecosystem can be explained in simple terms as a sustainable way of delivering ongoing innovation through the interaction of its elements.

3.1. Definitions

An innovation ecosystem refers to a loosely interconnected network of companies and other entities that coevolve capabilities around a shared set of technologies, knowledge and skills and work co-operatively and competitively to develop new products and services (Moore, 1993).

An innovation ecosystem is the evolving set of actors, activities, and artifacts, and the institutions and relations, including complementary and substitute relations, that are important for the innovative performance of an actor or a population of actors (Granstrand and Holgersson, 2020).

In this definition, artifacts include products and services, tangible and intangible resources, technological and non-technological resources, and other types of system inputs and outputs, including innovations. An innovation ecosystem could include an actor system with collaborative (complementary) and competitive (substitute) relations with or without a focal firm, and an artifact system with complementary and substitute relations.

An explicit systems approach to the studies of innovations was first adopted and developed in the economics and policy literature during the 1990s. A number of conceptualizations of innovation related systems were introduced, such as national, sectoral, regional and corporate innovation systems. The innovation ecosystem concept has partly come as a reaction to the value capture and competitive focus prevalent in the pre-existing business ecosystem literature, and that the innovation ecosystem concept put more emphasis on value creation and collaboration.

Box 3.1

One of the main actors in the mobile telecommunications ecosystem was Nokia, an active contributor to telecommunication standards and the world's largest mobile phone manufacturer during the first decade of the 2000s. However, after the introduction of Apple's iPhone, Nokia quickly lost its lead, which led its CEO Steve Elop to famously acknowledge: "Our competitors aren't taking our market share with devices; they are taking our market share with their entire ecosystem."

With the music player iPod, Apple had successfully positioned itself as a systems integrator and managed to build up an ecosystem of complementary technologies and actors, including content providers (record companies/ music publishers). Applying the same strategy in mobile telecommunications and smartphones proved to be extremely successful. By strictly controlling the smartphone hardware and operating system with a proprietary IP strategy while opening up for a plethora of complementary innovators in apps and contents, Apple managed to balance collaboration and competition, both regarding competition between itself and complementary actors (*e.g.*, in terms of how to split revenues from content sales), and competition between the different complementary actors (*e.g.*, between different gaming apps). Apple's innovation ecosystem allowed for successful complementors to reap sufficient, and sometimes huge, returns from their innovations, while at the same time allowing for competition among app providers, all in all leading to a dynamic innovation ecosystem in which complementary innovators were continuously providing new value to customers, part of which Apple could tap into and appropriate.

Source: Granstrand and Holgersson, 2020.

3.2. Functions of Innovation Ecosystem

Innovation ecosystems create an active flow of information and resources for ideas to transform into reality. Through these ecosystems, a process is built by which more innovators and entrepreneurs can develop and launch solutions to solve real-world problems, faster. This process creates expertise in new areas, helps to diversify the economy, and allows businesses to meet their customers where they are. Additionally, an innovation ecosystem provides the means to create economic stability and resource sharing.

Innovation ecosystems can operate at multiple levels (*e.g.* city, regional, national) and within multiple sectors (*e.g.* agriculture, health, education). Because of this range, it is difficult to draw meaningful boundaries around who is or is not part of an innovation ecosystem.

Adopting an ecosystems approach to innovation recognises that:

- ☆ An innovation ecosystem is made up of different actors, relationships and resources who all play a role in taking a great idea to transformative impact at scale.
- ☆ The effectiveness of each part within the innovation ecosystem is moderated by other parts of the system (*e.g.* entrepreneurs depend on being able to access financing).
- ☆ A change to one part of the innovation ecosystem leads to changes in other parts of the innovation ecosystem (*e.g.* an increase in internet connectivity will accelerate the design and testing of new technologies).

According to International Development Innovation Alliance (IDIA), "An Innovation Ecosystem is made up of enabling policies and regulations, accessibility of finance, informed human capital, supportive research, markets, energy, transport and communication infrastructure, a culture supportive of innovation and entrepreneurship, and networking assets, which together support productive relationship between different actors and other parts of the ecosystem".

3.3. Brief History of Innovation Policy in India

Since independence in 1947, India has been successful in building a massive ecosystem of science, technology and innovation (STI). An STI ecosystem includes universities, public and private enterprises, and human resources.

Five major policies related to technology and innovation have been implemented since independence till now, *viz.*, Scientific Policy Resolution (SPR 1958), Technology Policy Statement (TPS 1983), Science and Technology Policy 2003 (STP 2003), Science Technology Innovation Policy, 2013 (STIP 2013) and New STIP, 2020.

Box 3.2

Innovation and scaling require a commitment to doing things differently, and to find new ways of working together that transcend traditional boundaries to deliver greater impact, more quickly attained. In response to this challenge, The Rockefeller Foundation provided leadership and resources to convene some of the world's leading development innovation funders in early 2015 to form **The International Development Innovation Alliance (IDIA)**, a collaborative platform with the shared goal of *"actively promoting and advancing innovation as a means to help achieve sustainable development"*. IDIA is a unique collaboration platform that brings together the senior leadership from the innovation teams, labs and departments of some of the world's leading development agencies.

At the heart of this Alliance is a commitment to exchange innovation-related knowledge and learning across organisational, sectoral and geographic boundaries. IDIA seeks to achieve its Mission by sharing knowledge, experiences and insights; identifying and addressing gaps in the innovation ecosystem; and collaborating to develop public goods that support innovation, including pooling resources as appropriate.

Source: https://www.idiainnovation.org

SPR 1958

The first major science policy of India can be traced back to 1958. SPR 1958 recognised that science and technology were crucial for nation-building and national prosperity. It laid the foundation of scientific enterprise and scientific temper in India. SPR 1958 was primarily a science policy document. Technology development was given only secondary importance as it was assumed that technological development could only stem through the study of science and its applications.

TPS 1983

The primary feature of TPS 1983 was technological self-reliance through promotion and development of indigenous technologies. TPS 1983 recognised that though there should be efficient adoption of imported technologies, it should not be at the cost of national interest. Obsolete technologies had to be identified and replaced with technologies that can improve productivity, efficiency, quality and reliability of products using minimum capital and energy utilisation. To bolster indigenous technology development, the Technology Development Fund (TDF) was established to provide financial assistance to Indian industries.

STP 2003

SPR 1958 and TPS 1983 focussed on science and technology respectively and separately. However, the science and technology enterprises were so intertwined that there was a need to consider them together rather than as separate elements. STP 2003 called for cross-pollination of industry and scientific research, and the establishment of intellectual rights regimes to protect and incentivise inventors.

It was aimed at encouraging R&D and innovation in areas that impact the economy and society, and developing mechanisms to facilitate interaction among various stakeholders within the ecosystem. It also encouraged research in management and mitigation of natural hazards.

STIP 2013

The decade of 2010 to 2020 was declared as a decade of innovation by the then Prime Minister Dr. Manmohan Singh. It was acknowledged that in order to stay globally competitive, it was necessary to make a transition into a knowledge-based economy. STIP 2013 outlines the major policy initiatives to strengthen the innovation ecosystem and give a boost to the development of innovation-led entrepreneurship in India. Promoting a science and technology-led innovation ecosystem in the country and broadly linking science, technology and innovation to socio-economic priorities were some of the key aspects of STIP 2013. This policy document was a step in the right direction towards building a robust national innovation ecosystem.

The Policy tried to join the fragmented pieces of the Indian innovation ecosystem and bring it into the sharper focus. It addressed the need to enhance scientific understanding and skills among the young and expected to position India among the top five global scientific powers by 2020. It also linked the contributions of science, research, and innovation with an inclusive growth agenda with the aim of forming a robust and focused national innovation system. The policy supported entrepreneurship driven by science, technology, and innovation with viable and highly scalable business models. A key mechanism was investment in young innovators and entrepreneurs through education, training, and mentoring.

Present Scenario

The new and recent Science, Technology, Innovation Policy of 2020 aims to bring about profound changes through short-term, medium-term, and long-term mission mode projects by building a nurtured ecosystem that promotes research and innovation on the part of both individuals and organizations.

The policy envisions strengthening of the overall innovative ecosystem, fostering Science and Technology (S&T)- enabled entrepreneurship, and improving participation of the grassroots levels in the research and innovation ecosystem. A participative model with four interconnected tracks has been envisioned under the STIP 2020.

1. **Track I Extended public and expert consultation** aims to capture the aspirations of a larger set of stakeholders and create a repository of public voices that will act as a guiding force for the drafting process. There are six unique activities under this track (live virtual conversations with thought leaders, thematic public panel discussions, focused survey instruments, digital and print media campaigns, community radio broadcasts and a competition called 'the ideathon' to ignite minds), designed carefully keeping the limitations (access, reach, digital services, language barriers and last-mile connectivity) of different stakeholder groups in mind. Track I, through the Science Policy Forum, engages with a larger pool of public and experts, and gathers inputs during and after the policy drafting process.
2. **Track II Thematic Groups** comprise experts-driven thematic consultations to feed evidence-informed recommendations into the policy drafting process. Twenty-one focused thematic groups have been constituted for this purpose.
3. **Track III Ministerial and state consultation** involves consultations with Ministries and States for which nodal officers are nominated in States and in Ministries, Departments and Agencies of Government of India for extensive intra-state and intra-department consultation. The 'State/UT Consultations' focus on strengthening the regional STI ecosystem and creating efficient means of STI Policy Governance at the state level that can lead to the creation of state STI Policies and action plans in tandem with the national STIP 2020. The ministerial consultation process draw key policy recommendations from different ministries and incorporate them into a form that can lead to the creation of a broader STI policy governance and robust STI data architecture.
4. **Track IV Apex Level Multi-stakeholder Engagement-** This track is the binding force that draws upon the apex-level multi-stakeholder engagement at the national as well as global levels. This track is ex-officio in nature, involving representatives from governments, academia, industry and civil society organizations. The Track-IV apex-level consultations have focused deliberations planned with (i) young scientists and technologists, (ii) civil society organizations with special focus on farmers, (iii) scientific ministries, departments and agencies, (iv) socio-economic ministries and department, (v) state governments, and (vi) global partners.

The policy will promote technology self-reliance and indigenization to achieve the larger goal of "Atmanirbhar Bharat". A two-way approach of indigenous development of technology as well as technology indigenization will be adopted and focused upon in alignment with national priorities, like sustainability and social benefit, and resources. International engagements will be facilitated to gain essential know-how towards creation and development of indigenous technologies. The chapter 'Technology Development and Indigenisation' outlines strategies for strengthening India's STI ecosystem to achieve the larger goal of "Atmanirbhar Bharat" through technology self-reliance and indigenisation.

A dedicated portal, Indian Science and Technology Archive of Research (INDSTA), will be developed to provide access, specifically, to the outputs of all publicly-funded research. The new policy has a chapter on 'Innovation and Entrepreneurship' which aims at creating an enabling ecosystem for seeding, sustenance, and growth of science and technology entrepreneurship in India. This chapter outlines strategies for strengthening the overall innovative ecosystem, fostering S&T-enabled entrepreneurship and improving participation of the grassroots levels in the research and innovation ecosystem.

S&T-enabled entrepreneurship-technology-driven, innovation-focused enterprises is critical for creating, shaping, and sustaining the future industrial sectors of the country as well as delivering the benefits of scientific research for socio-economic development of the society.

India needs to develop and deploy sustainable technologies at a faster rate to successfully tackle the major socio-economic challenges and changing aspirations of the people. Along with this, India has to develop and deploy those technologies indigenously which aim for bringing sustainability and inclusive growth benefiting the society at large.

3.4. Innovation Ecosystem in India

Broadly speaking, an innovation ecosystem is a combination of two distinct but largely separated economies: i) the knowledge economy (comprised of knowledge producers), driven by fundamental research, and ii) the commercial economy (comprised of knowledge users), driven by the marketplace (Jackson, 2013).

An important feature of an innovation ecosystem is that the resources available to the research economy are coupled to the resources generated by the commercial economy, usually as some fraction of the profits in the commercial economy. Another feature is that entities within the ecosystem are either geographically localized or strategically linked to focus on developing a specific technology. Silicon Valley is the best known example of a geographically localized ecosystem.

In India, the innovation ecosystem includes the entire national innovation system plus individual innovators and entrepreneurs; mentors; government policies; angel, venture capital, institutional, and industrial funding mechanisms, intellectual property rights mechanisms; technology transfer mechanisms; market inputs; and incentives, awards, and other innovation-recognition mechanisms, among others. Ideally, these various structures and mechanisms facilitate the smooth translation of innovations through the various segments of a complex innovation chain that takes ideas from "mind to market".

The current national innovation system in India is a vast and complex system comprising of knowledge producers such as science and technology institutions, academia, and innovating individuals and knowledge users (*e.g.*, industry-production/services in the public and private sectors). Various governments in India have given priority to science, technology, and innovation, and therefore India has evolved a large publicly funded R&D structure. There are various councils and research structures under different ministries, which cater to different research areas and distributed across the country. These institutions were working mostly in a standalone mode prior to linearization of the economy in the 1990s. Since then, strong efforts have been made to harness the innovative capabilities of these structures by connecting them to one another and to industry and society, forming an innovation ecosystem

In spite of a large publicly funded science and technology infrastructure and a sizeable education base, India has not been able to realize its full innovative potential due to a fragmented innovation ecosystem. Realising this, the Government of India has taken many initiatives towards strengthening the innovation ecosystem.

The two most important earlier initiatives were- i) the establishment of the National Innovation Council, whose mandate is to coordinate various innovation-related activities, and ii) the Science, Technology and Innovation Policy 2013, intended to promote entrepreneurship and science-led solutions for sustainable and inclusive growth.

National Innovation Council

The National Innovation Council (NInC) is the think tank council set up under the Chairmanship of Mr. Sam Pitroda to discuss, analyse and help implement strategies for inclusive innovation in India and prepare a Roadmap for Innovation 2010-2020. NInC is considered the first step in creating a crosscutting system which will provide mutually reinforcing policies, recommendations and methodologies to implement and boost innovation performance in the country.

The vision of NInC was to reorganize the Indian model of innovation from a culture of ad hoc innovation, or "Jugaad", driven by scarce resources and customers' needs, to a system based on strategic approaches that incorporate sustainability, durability, affordability, quality, global competitiveness and local needs.

The NInC was established as the central hub to administer government activity in support of innovation, providing assistance via human, financial, social and intellectual capital to the following components: State, Sectoral and City Innovation Councils; Cluster Innovation Centres; India Inclusive

Innovation Fund; 'One MP-One Idea' initiative, Innovation in Education and Global Innovation Roundtable.

The NInC's strategy is to act as a "catalysing agent" to bring about a long-term cultural change. Rather than independently implementing initiatives, the NInC acts as a consultant and facilitator, reducing bureaucratic bottlenecks and bridging market disconnects.

The tasks of the National Innovation Council, State Innovation Council and Sectoral Innovation Council are given in **Annexure-2**

India Inclusive Innovation Fund

The Ministry of Micro Small and Medium Enterprises (MSME), on the advice of the erstwhile National Innovation Council, has set up a dedicated fund in the name of 'India Inclusive Innovation Fund' for promoting grassroot innovations with social returns as well as modest economic returns. The India Inclusive Innovation Fund (IIIF) is designed to "combine innovation and dynamism of enterprises to solve the problems of the bottom of the pyramid in India" (National Innovation Council, 2014).

This fund would, operate as a for-profit entity with a social investment focus. The Fund would back enterprises developing innovative solutions primarily for citizens who lie in the lower half of India's economic pyramid, with limited physical and institutional access to basic services.

The total corpus of the fund is proposed to be a minimum size of Rs. 500 crore and maximum size of Rs.5,000 crores with initial Government of India contribution of Rs. 100 crore and the rest to be obtained from financial institutions, insurance companies, multilateral/bilateral development agencies, and Indian and global corporations. The life of the fund is nine years, and it will focus on healthcare, food and nutrition, agriculture, education energy, financial inclusion, and environment technology, among other areas.

Since National Innovation Council has ceased to exist, a revision in the three level of Management structure has been proposed. The Fund would operate as a for-profit entity with a social investment focus. The exact extent to which the India Inclusive Innovation Fund is likely to benefit the MSME Sector has not been estimated. However, it is proposed to back enterprises developing innovative solutions primarily for citizens who lie in the lower half of India's economic pyramid, with a limited physical and institutional access to basic services.

3.5. Key Actors in Indian Innovation Ecosystem

The Ministry of Science and Technology (S&T) is the key Ministry which plays a pivotal role in the promotion of science and technology in the country. Technology Development Programme (TDP) is one of the major activities of the Ministry. The Department of Science and Technology (DST) and the Department of Biotechnology (DBT) are two major departments of the Ministry.

3.5.1. Department of Science and Technology (DST)

Technology Development Programmes

One of the key objectives of the Ministry is to promote technology development in various fields. DST has been supporting Technology development projects which include materials, devices and processes. The Programme supports activities aimed at developing technologies both in the advanced/emerging areas and in traditional sectors/areas. Under the programme, feasibility of fresh ideas/concepts is also assessed for their potential conversion into useful technology/product.

The mandate of TDP is to convert proof-of-concepts for technologies/ techniques/processes/products into advance prototypes for validation and demonstration in actual field settings. The commercialization of these technologies needs further assessment/incubation, which does not fall in the scope of the Technology Development Programme. Transfer of technology developed under the project to the industry is the responsibility of the host institutions. Proposals for incremental R&D over the existing technologies are also considered for support. Projects related to design and development of software/IT, as required for products and processes, as part of technology development project are also considered. However, pure software development does not fall in the scope of the Programme.

The India Innovation Growth Programme (IIGP)

It is a unique tripartite initiative of the DST, Government of India, Lockheed Martin and Tata Trusts. It is aimed at enhancing the Indian innovation ecosystem by enabling innovators and entrepreneurs through the stages of ideation, innovation and acceleration, to develop technology-based solutions for tomorrow. The programme aims to build an innovation pipeline in India through a high-impact programme focused on the social and industrial innovation ecosystem.

Lockheed Martin Corporation is an American aerospace, arms, defence, information security, and technology corporation with worldwide interests. With a presence in India for over 30 years, Lockheed Martin opened India subsidiary in New Delhi in 2008 and has supported and aligned itself with various initiatives of the Government of India like Make in India, Startup India and Skill India.

Supporting the Government of India's missions of "Start-up India" and "Make in India", IIGP enhances the Indian innovation ecosystem by enabling innovators and entrepreneurs through the stages of ideation and innovation, to develop technology-based solutions for the future. Complemented by several implementation partners, Federation of Indian Chambers of Commerce and Industry (FICCI), Indo-US Science and Technology Forum (IUSSTF), Centre for

Innovation Incubation and Entrepreneurship (CIIE) at IIM Ahmedabad, Indian Institute of Technology Bombay and the Tata Centre for Technology and Design at the Massachusetts Institute of Technology (MIT), the programme aims to build an innovation pipeline in India through a high-impact programme focused on the social and industrial innovation ecosystem.

Launched in 2007, IIGP has been one of India's longest standing public-private partners. The programme has provided mentoring and handholding assistance to over 400 innovators from diverse sectors from across the country and generated over 350 commercial agreements.

Technology Information Forecasting and Assessment Council (TIFAC)

TIFAC is an autonomous organization set up in 1988 under the DST to look ahead in technology domain, assess the technology trajectories, and support innovation by networked actions in select areas of national importance. TIFAC continues to strive for technology development in the country by leveraging technology innovation through sustained and concerted programmes in close association with industry and academia. The journey of TIFAC has been characterized by a mix of technology foresight and innovation support activities, spanning from individual innovator to small and medium scale enterprises to pre-competitive technology development efforts in consortia mode.

National Initiative for Developing and Harnessing Innovations (NIDHI)

It is an umbrella programme conceived and developed by the Innovation and Entrepreneurship division of DST for nurturing ideas and innovations (knowledge-based and technology-driven) into successful startups. The programme works in line with the national priorities and goals and its focus would be to build an innovation driven entrepreneurial ecosystem with an objective of socioeconomic development through wealth and job creation.

NIDHI aims to nurture start-ups through scouting, supporting and scaling of innovations. The key stakeholders of NIDHI includes various departments and ministries of the central government, state governments, academic and R&D institutions, mentors, financial institutions, angel investors, venture capitalists and private sectors.

DST is supporting Technology Business Incubators primarily in and around academic, technical and management institutions to tap innovations and technologies for venture creation by utilizing expertise and infrastructure already available with the host institution.

Technology Business Incubator (TBI)

It is an organisational setup that nurtures technology based and knowledge driven companies. They are a venture of universities, public research institutes, local government and private players to promote and bolster a new technology intensive enterprise.

TBIs provide budding entrepreneurs all necessary infrastructure support, technology/prototype development support, research assistance, help in getting funding, business consulting assistance, marketing assistance and undertake whatever is necessary to make the start-up a success. The main objective of the technology business incubators is to produce successful business ventures that create jobs and wealth in the region, along with encouraging an attitude of innovation in the country as a whole.

National Innovation Foundation (NIF)

NIF is an initiative to strengthen the grassroots technological innovations and outstanding traditional knowledge. Its mission is to help India become a creative and knowledge-based society by expanding policy and institutional space for grassroots technological innovators. It was set up in March 2000 with the assistance of DST, Government of India

NIF scouts, supports and spawns grassroots innovations developed by individuals and local communities in any technological field, helping in human survival without any help from formal sector. NIF helps grassroots innovators and well known traditional knowledge holders get due recognition, respect and reward for their valuable innovations. It also tries to ensure that such innovations diffuse widely through commercial and/or non-commercial channels, generating material or non-material incentives for them and others involved in the value chain.

NIF has also set up an augmented Fabrication Laboratory (Fab Lab) for product development and strengthening in-house research. Micro Venture Innovation Fund (MVIF) at NIF, with the support from Small Industries Development Bank of India (SIDBI), has provided risk capital to many innovation based enterprise projects, some of which are at different stages of incubation. NIF has also succeeded in commercialising products across countries in six continents, apart from being successful in materialising many cases of technology licensing.

The INSPIRE Award- MANAK (Million Minds Augmenting National Aspiration and Knowledge) is being revamped and executed by the DST and NIF-India to align with the action plan for 'Start-up India" initiative launched by the Government of India.

National Innovation Portal (NIP) developed by NIF- India is currently home to about 1.15 lakh innovations scouted from common people of the country, covering Engineering, Agriculture, Veterinary and Human Health. In terms of domain areas, presently the innovations cover Energy, mechanical, automobile, electrical, electronics, household, chemical, civil, textiles, Farm/cultivation practice, storage practice, plant variety, plant protection, poultry, livestock management, nutraceuticals *etc.*

3.5.2. Promoting Innovations in Individuals, Startups and MSMEs (PRISM)

It is an initiative of the Department of Scientific and Industrial Research (DSIR) aimed at transforming an individual innovator into a successful technopreneur by promoting, supporting, and funding implementable and commercially viable innovations created. DSIR is a part of Ministry of S&T which has a mandate to carry out the activities related to indigenous technology promotion, development, utilisation and transfer.

PRISM aims to transform an individual innovator into a successful Technopreneur by promoting, supporting and funding the untapped creativity of individual innovators. It is meant for budding students, professionals and common citizens interested in innovation and having an implementable and commercially viable, novel innovation.

3.5.3. Innovation Management Directorate of CSIR

Council of Scientific and Industrial Research (CSIR) under the Ministry of S&T, a premier national R&D organisation, is among the world's largest publicly funded R&D organisation. The R&D efforts of CSIR are aligned to the needs of the domestic industry and the social needs of the nation exemplified by the Sustainable Development Goals (SDGs) chartered by UN as well as the policies and vision of the government.

The R&D of CSIR cater to the Government initiatives such as Make in India; Swachch Bharat; Clean Ganga; Swaasthya Bharat; Smart Cities; Smart Villages; Clean, efficient, affordable and renewable energy technology solutions; Innovate in India; *etc.* in a significant way.

Innovation Management Directorate in CSIR was created in May 2020 with the mandate of

- ☆ Leveraging Science, Technology and Innovation (STI) for meeting the needs of the country and economic value creation: Focused, Benchmarked R&D and Technology Development efforts through CSIR Mission-mode Projects,
- ☆ Enhancing Industry Partnerships: Through public private partnership projects under CSIR-NMITLI with the objective to accelerate technology deployment and commercialization at large scale.

New Millennium Indian Technology Leadership Initiative (NMITLI) is the largest public-private-partnership effort within the R&D domain in the country. NMITLI has carved out a unique niche in the innovation space and enjoys an excellent reputation. NMITLI has so far evolved 60 largely networked projects in diverse areas including Agriculture and Plant Biotechnology.

3.5.4. Department of Biotechnology (DBT)

Since its inception in 1986, DBT has been committed to delivering solutions for societal challenges with a focus on 'Make in India'. It has been a catalyst in spearheading innovative solutions for developmental challenges. Biotechnology today has played a major role in impacting food and agriculture, nutrition, healthcare, the environment and industrial growth. As the field has progressed, DBT has made significant strides and aided in the economic and social growth of the country.

Over the last three decades, DBT has created a very strong research and translation ecosystem across the country and has built strong foundations, leveraging the strength of national and international partnerships. With more than 15,000 scientists and 800 institutes and laboratories supported, DBT supports nearly 10,000 biotechnology research fellows and students annually. World class state of the art infrastructure has been created, which through the DBT's SAHAJ scheme has now been made accessible to all researchers and startups to take research and innovation across the country. SAHAJ refers to Scientific Infrastructure Access for Harnessing Academic University Research Joint Collaboration.

Skill Vigyan life science and biotechnology centres have helped build an employable skilled human resource base. DBT has also made significant contributions by aligning its work to the national growth agenda and developing innovative solutions for the national missions of Swasth Bharat, Ayushman Bharat, Swatch Bharat, Poshaan Abhiyan, Start-up India, Make in India and Skill India. Academic and industry experts have successfully delivered to the country.

Biotechnology Industry Research Assistance Council (BIRAC)

It is a not-for-profit Section 8, Schedule B, Public Sector Enterprise, set up by the DBT as an Interface Agency to strengthen and empower the emerging Biotech enterprise to undertake strategic research and innovation, addressing nationally relevant product development needs. BIRAC is an industry-academia interface and implements its mandate through a wide range of impact initiatives.

It was set up to promote and nurture a vibrant startup and entrepreneurial ecosystem. BIRAC aims to foster excellence while supporting bio-innovations for societal benefit. To take scientific research from the laboratory to the end user, the DBT has built an ecosystem that allows for seamless movement of research from the translation phase to commercialisation

The DBT has consistently brought to the fore the importance of science in all walks of life and has integrated innovation with national development. It has promoted science and technology as a way of thinking rather than a siloed field of work and envisions to continue to do so, and be a critical part of India's growth story.

Centres of Excellence and Innovation in Biotechnology (CEIB)

Aiming to provide funding to augment and strengthen institutional research capacity for promotion of excellence in interdisciplinary science and innovation in specific areas of biotechnology, the DBT has launched a group of schemes known as CEIB.

The overall aim of the CEIB programme is to establish Centres of Excellence, provide long-term support to outstanding scientists and provide institutional/ departmental programme support. The specific goal is to enhance the innovative ability of the institutions and investigators with well-developed research programme in specific areas of biotechnology.

The programme is intended for institutions with a substantial investment in, and commitment to, life sciences and biotechnology research. The programme will provide flexible long-term support for highly innovative research in life sciences and biotechnology. The expected outcomes are quality science, intellectual property and potential translation.

Small Business Innovation Research Initiative (SBIRI)

SBIRI of the DBT was launched in 2005 to boost Public-Private- Partnership (PPP) efforts in the country. SBIRI was the first of its kind, early stage, innovation focused PPP initiative in the area of Biotechnology. Launching of SBIRI has worked as an enabling platform for the target organizations to realize their potential in terms of product and process development and taking them to the market. It has facilitated innovation, risk taking by small and medium companies and bringing together the private industry, public institutions and the government under one roof to promote the research and innovation in the Indian Biotech Sector.

3.6. Innovation Ecosystem in Agriculture Sector

With the need to allow agricultural systems to remain competitive in a changing socioeconomic environment, the concept of agricultural innovation system (AIS) was merged to reflect the importance of going beyond the creation of knowledge and highlights that innovation systems are social systems. The Agricultural Innovation System is a school of thought that emphasizes learning about how innovation can be scaled up and innovation capacity can be developed among stakeholders. Innovation systems is a framework to better understand the process of knowledge generation, sharing, access and exchange among stakeholders and put them into social and economic use. Recently, the conceptualization of agricultural innovation systems to better support sustainability is called Agricultural knowledge and innovation system (AKIS).

AKIS has a strong potential to enhance economic performance of farming and contribute to agricultural sustainability, as it may increase synergies and complementarity among actors. AKIS is broadly defined as a network of individuals, enterprises, and organizations focused on developing innovations with regard to new products, new processes, and new forms of organization into economic use, together with policies and institutions that facilitate the way different stakeholders interact, access, share, exchange, and use knowledge. This framework includes three essential elements: (a) A knowledge and education domain; (b) A business and enterprise domain; and (c) Linkages between institutions that link the two domains.

AKIS indicates a system that links people and institutions to promote mutual learning. While AKIS focussed on the knowledge and innovation processes in the rural sphere, AIS focussed more on building innovation capacities of stakeholders in a multi-stakeholder process with reduced focus on public research system for knowledge.

During 1980s and 1990s, research was a major priority with central focus and budgetary allocation favouring setting up of national research institutes. In the later part of 1990s, the focus gradually moved from research institutes to more broad-based, pluralistic National Agricultural Research Systems (NARS).

NARS focusses on the generation of knowledge, AKIS on the generation and diffusion of knowledge, and AIS on the generation, diffusion, and application of knowledge (Roseboom, 2011).

3.6.1. Indian Council of Agricultural Research (ICAR)

ICAR is the apex body for coordinating, guiding and managing research and education in agriculture including horticulture, fisheries and animal sciences in India. It is an autonomous organisation under Department of Agriculture Research and Education (DARE), Ministry of Agriculture and Farmers' Welfare, Govt. of India. The ICAR has played a pioneering role in ushering Green Revolution and subsequent developments in agriculture in India through its research and technology development that has enabled the country to ensure food security of the country.

ICAR's three tier system of technology commercialization process effective since 2007, has Intellectual Property and Technology Management (IP and TM) Unit at the HQ as the top tier, Zonal Technology Management Centres (ZTMCs) as the middle tier and Institute Technology Management Units (ITMUs) at the bottom level for effective technology transfer through licensing.

IP and TM Unit

The Intellectual Property and Technology Management Unit in ICAR oversees all matters related to intellectual properties and technology transfer/ commercialization. ICAR recognizes that a systematic management of its technology products and services while bringing commercial ethos in their transfer and realization at the user end would result in much-needed dividends for the nation. Individual institutes of ICAR are empowered and enabled to enter into licensing contracts or commercial agreements for the commercial transfer of ICAR technologies to the interested parties.

Assimilation of IPR dimensions in research management has become absolutely necessary in R&D institutions. Accordingly, ICAR has formulated Guidelines for Intellectual Property Management and Technology. A decentralized three-tier IP management mechanism is also institutionalized in ICAR from 2006 onwards.

A scheme titled. "Intellectual Property Management and Transfer/ Commercialization of Agricultural Technology" has contributed to developing an IP environment in ICAR. Its implementation resulted in increased IPR-filing (patents, plant variety rights, copyrights, trademarks, *etc.*).

A middle-tier consisting of Ten Subject Specific Zonal Technology Management (ZTM) Units is developed to facilitate serious business and strengthen public-private partnerships. These zonal units also project the ICAR technology profiles available at various ICAR institutes located in the respective zones. The central IP and technology management mechanism at ICAR Headquarters facilitates techno-legal and policy matters/concerns that may arise on case-to-case basis in the course of developing win-win relationships and also catalyzes public-private relationships at the central level.

Policy and Guidelines

ICAR encourages interested parties to enter into commercial ventures with ICAR technology and knowledge products and services as per the:

- ☆ ICAR Guidelines for Intellectual Property Management and Technology Transfer/Commercialization (Revised 2018)
- ☆ ICAR Rules and Guidelines for Professional Service Functions (Training, Consultancy, Contract Research and Contract Service)
- ☆ Empanelment of law firms for Patent - IPR Attorneys
- ☆ Agri-Startups: Reflection of ICAR Technologies in Market
- ☆ Proceedings of UPAYA (Agri-Startups and entrepreneurship Conclave)

3.6.2. National Agriculture Innovation Fund

Based on the encouraging lessons and considering the enhanced scope, a scheme titled '***National Agriculture Innovation Fund***' with three components is being implemented by ICAR. The components are: Component I: Innovation Fund, Component II: Incubation Fund (Supporting Agri-business Incubation Centres in institutions developing agricultural technologies); and Component III: ARYA (Attracting and Retaining Youth in Agriculture).

The objectives of the three components are:

Component I (Innovation Fund)

1. To promote creativity and innovations in ICAR institutes;
2. To strengthen the institutional mechanism to protect/manage innovations/intellectual properties (IPs) generated within the ICAR/NARS system;
3. To understand and attempt as necessary a vigil on infringement, improper exploitation and abuse of the intellectual assets belonging to ICAR/NARS;
4. To augment capacity building in IPRs and technology transfer/commercialization dimensions;
5. To manage new knowledge as per principles in ICAR's new and implemented guidelines/policies.

ICAR has planned a system that will scientifically authenticate, scale up, and support the innovations of reformist farmers. It will set a centre for innovation in New Delhi to validate their innovations scientifically and allow them to pursue research work. The system will link farmers and farming with science and encourage them to continue their innovations.

Component II (Incubation Fund)

1. To establish/transform agri-business Incubator centres as leaders in NARS that would provide technology and skill up gradation, inputs supply and market support leading to promotion of viable enterprises and sustainable employment to entrepreneurs;
2. To undertake last mile scale-up from pilot level of value chain in collaboration with stakeholders;
3. To explore and support appropriate technologies including grassroots' innovations that are vital in future for an accelerated growth and competitive technological leadership;
4. To impart training and capacity building to prospective entrepreneurs; generate value added manpower to compete effectively; and
5. To provide seed money support to potential incubates/entrepreneurs taking up promising innovations or technologies.

ICAR emphasizes that agriculture is a science, and scientific techniques can boost agriculture. It has formed a tie between 105 startups and farmers to promote technology in the farming sector.

Component III (ARYA)

1. To attract and empower the Youth in Rural Areas to take up various Agriculture, allied and service sector enterprises for sustainable income and gainful employment in selected districts.

2. To enable the Farm Youth to establish network groups to take up resource and capital intensive activities like processing, value addition and marketing.
3. To demonstrate functional linkage with different institutions and stakeholders for convergence of opportunities available under various schemes/program for sustainable development of youth.

3.6.3. Agrinnovate India

Realizing the need for a 'corporate face' that could promote technology transfers in a centralized manner, unifying the processes across its research institutions, Agrinnovate India Limited (AgIn), a Government of India Enterprise was established in October, 2011 under the Companies Act, 1956 (No. 1 of 1956) under the direct administrative control of DARE, Ministry of Agriculture and Farmers' Welfare, Govt. of India.

It is a "for profit" Company owned by DARE. It is to act as an effective interface between Indian Council of Agricultural Research on one side and the Stakeholders of agricultural sector (Farmers; Public and Private Sector firms; R&D organizations; Educational Institutions- all of these at National and International level) on the other side, for a significant purpose of securing, sustaining and promoting global agricultural development.

The main objectives of Agrinnovate India are:

- ☆ Protection, management, commercialization and distribution of Intellectual Property for Public Benefit.
- ☆ Production, marketing and popularization of ICAR's products, processes and technologies in agriculture and allied sectors such as seed, planting material, vaccines, diagnostics, biotechnological products, other value added inputs and products and farm implements and machinery.
- ☆ Providing consultancies, contract research, contract service, customized capacity building.
- ☆ Setting up of research and production farms outside India and undertaking global brand building initiatives.
- ☆ Providing technical support for turnkey projects on production and processing plants in agriculture and allied sectors.
- ☆ Creation of public-private partnerships in research, education and other capacity building endeavours in agriculture and allied sectors.
- ☆ Carrying out activities to integrate proficiencies in agricultural sciences with management, such as market intelligence, pricing and valuation issues, to nurture demand-driven research.

Box 3.3

Banana Pseudo stem sap, a by-product of extraction of banana fibre, is a rich source of plant nutrients and growth regulators. When enriched with organic inputs through anaerobic incubation, the sap acts as a rich source of nutrients replacing the use of chemical fertilizers and is also suitable for use in organic farming system. Agrinnovate India Limited facilitated the tech transfer and establishing an International linkage with Ms. Accrued Gains Pty. Ltd., Republic of Botswana, by licensing 'Banana Pseudo stem sap for use as biofertilizer for crop productivity enhancement. The technology developed by Navsari Agricultural University (NAU), Gujarat was transferred through a commercial non-exclusive license for a period of five years (December 2019).

Source: Annual Report 2019-20, Agrinnovate India Private Limited.

3.7. Enabling Environment for Innovations in Agriculture

The 'enabling environment' for agricultural innovation encompasses factors that influence agricultural innovation positively but are controlled by policy domains other than agricultural innovation policy. An agricultural innovation policy seeks coordination with other domains to ensure that together they enable agricultural innovation. Cross-cutting policy issues affecting agricultural innovation include policies to reduce poverty and sustain the environment, to foster collaboration between the public and private sectors, and to build social capital more generally.

Three clusters of enabling factors for agricultural innovation appear to require attention and investment in most developing countries: (1) innovation policy and corresponding governance structures to strengthen the broader framework for agricultural innovation policies; (2) regulatory frameworks that stimulate innovation directly (such as IPRs) or indirectly (standards that stimulate trade) or steer innovation towards certain preferred outcomes (safer food); and (3) accompanying agricultural investments in rural credit, infrastructure, and markets.

Innovation policy is a new area, and in most countries the governance structure for innovation is only starting to emerge. A particular challenge is where to assign responsibility for innovation policy within the government structure. Some countries delegate this task to the ministry in charge of science and technology, while others establish a higher-level entity that brings relevant ministries together to coordinate national innovation policy. In most countries, the overall objective of the national innovation policy is to facilitate the transition toward a knowledge economy, resulting in increased competitiveness and sustainable economic growth.

Investment in a national innovation policy and corresponding governance structures strengthens coordination across policy domains on innovation issues, addresses issues relevant across sectors (such as IPRs or tax deductions for innovation), provides a framework for more sector-specific innovation policies (including an agricultural innovation policy), and, prioritizes public innovation investments across sectors. In many instances, national innovation policy has generated a substantial influx of new ideas and instruments into the agricultural innovation domain, including such concepts as business incubators and risk capital.

In agriculture, India developed a national agricultural research system and an agricultural extension system alongside the Green Revolution that significantly increased agricultural productivity and helped increase the diversity of India's agriculture.

The onset of economic reform introduced a series of significant shifts in the policy framework for science, technology and innovation. The initial thrust of the reform rhetoric was focused more on dismantling of the regime of controls and licensing and its implications for technology development and innovation. The second key shift was the increasing acceptance of the global intellectual property rights regime. The third key shift was the emphasis on the private sector as the key sector to drive innovation forward.

The U.S.- India Knowledge Initiative on Agriculture (KIA) launched in 2005 by President Bush and Indian Prime Minister Dr. Manmohan Singh aimed to strengthen the Indian agricultural knowledge system in areas of mutual interest. With the involvement of universities and private sector, the initiative was focused on building human and institutional capacity, promoting sustainable use of water resources, effectively applying biotechnological approaches, and ensuring strong markets and processing chains. However, the design of the Indo–US engagement in agriculture during both the Bush and Obama Administrations was considered problematic and the KIA focused its attention only on creating a favourable climate for contract farming in India.

The Knowledge Initiative in Agriculture to promote collaboration in agriculture, is considered by the Indian Government as the second Green Revolution. Its hallmark is that it deals with the transfer of proprietary knowledge in agriculture, including GM technology, mostly through MNCs and not through public institutions. The contrasts in the KIA with the original Green Revolution are striking.

The Patent and Trademark Law Amendments Act (known popularly as the Bayh–Dole Act), enacted by the US Congress in 1980, allows inventions arising from publicly funded programmes to be patented. As a result, public institutions do not only imitate the private entities in the US; they have also actually been licensing patents arising from public funding to other private businesses. This was considered a threat to scientific technologies developed in India.

NIFA (National Institute of Food and Agriculture) under USDA also supported the Structuring Agricultural Marketing Systems project in India through collaboration with India's National Institute of Agricultural Marketing (NIAM), responsible for agricultural marketing-related promotion and training. In partnership with USDA's Foreign Agricultural Service and Agricultural Marketing Service, NIFA provided technical assistance to NIAM and state agencies in the development of Good Agricultural Practices, post-harvest handling techniques and marketing extension. India's existing marketing news information and marketing extension system was also upgraded to better serve farmers. This was accomplished through the development and implementation of train-the-trainer workshops for NIAM and other government staff conducted by experts from several U.S. land-grant universities.

3.8. Innovations for Rural India

3.8.1. Accelerating Growth of New India's Innovations (AGNIi)

It is a programme of the Office of the Principal Scientific Adviser to the Government of India, and a Mission under the Prime Minister's Science, Technology and Innovation Advisory Council (PM-STIAC). The aim of AGNIi is to support the national efforts to boost the innovation ecosystem in the country by connecting innovators across industry, individuals, and the grassroots to the market and helping commercialise innovative solutions. It provides a platform for innovators to bring their technology-ready products and solutions to industry and the market, thereby helping propel techno-entrepreneurship which can usher in a new era of inclusive socio-economic growth.

AGNIi helps commercialise Indian technological innovation. The core mission of AGNIi is to support and augment initiatives undertaken by multiple stakeholders driving India's innovation goals and help connect owners of innovative and new-frontier solutions with the market. It convenes India's finest technologists and their deployment-ready technologies- from startups and laboratories- with top technology users from Indian and global enterprise, government and non-profit sectors.

The AGNIi Mission, in partnership with Common Service Centre, successfully conducted a series of Digital Technology Showcases focused on innovations for rural India. Recently, in August 2021, two 90-minute showcases were held in Bihar, to address the following concerns:

- ☆ Drinking water issues due to contamination of water or lack of proper storage
- ☆ Soil testing technologies with less technical dependency to ensure more farmers can use them easily
- ☆ Storage technologies for vegetables and fruits

☆ Drones for spraying of pesticides as due to land distribution in all major districts it becomes difficult to take heavy machinery from one farm to the other

Seven technologies were presented during the showcases reaffirming the capability of Indian innovations to tackle challenges faced by Indian villages:

1. **Dharashakti:** An easy-to-use soil organic carbon detection and testing kit
2. **Tan 90:** Portable cold storage units to plug the gaps in cold chain
3. **Krishakti:** Smart and intelligent precision spraying drones for crops
4. **Avatar Small Wind Turbine:** Wind turbines deployed as decentralized and distributed renewable energy solutions for urban and rural India
5. **Saptkrishi:** A low-cost, technological solution that extends the shelf life of perishable horticultural produce.
6. **Soil Saathi:** A lightweight chemistry analyzer with the ability to test 22 parameters in soil and water and advise fertilizer requirements based on crop and agro-climatic conditions.
7. **Thanos:** A drone-based autonomous spraying platform that sprays an acre in a fraction of the time uniformly.

The AGNIi Mission remains committed to introducing emerging technologies in rural India to address the complex and unique challenges it faces today.

3.8.2. Atal Innovation Mission (AIM)

AIM is Government of India's flagship initiative to promote a culture of innovation and entrepreneurship in the country. Creating a nation of job-creators and not just job-seekers is the focus of this mission. The need for an extensive collaboration between corporate industry, academia, and governments at the village, district, state and central levels is also highlighted.

There are five pillars of AIM which can be viewed through the lens of innovation and entrepreneurship.

1. **Pillar of demographic dividend:** The channelisation of the youthful energy towards nation-building activities has to be triggered by focusing on developing vocational, technical and managerial skills, while fostering a culture of innovation and entrepreneurship at the school, university and industry levels.

   Innovative ways of leveraging emerging digital technologies can create and promote such ecosystems. Existing government initiatives such as Atal Tinkering Labs and AI curricula at school levels, incubators and innovation cells at university levels, and fostering startups and vocational training in emerging new areas can also be accelerated to strengthen this pillar.

2. **Pillar of infrastructure**: The nation needs smart villages and several hundred smart cities to become active hubs of livelihood enablement, innovation, and job creation. The pillar of infrastructure would also require construction of digital highways, which would ensure that innovations reach every common citizen be it in education, healthcare, housing, or job enablement.
3. **Pillar of Demand:** India has the perfect environment to meet the demand. There is an ideal situation for thousands of startups and companies to capitalise on and fulfil the demand for new solutions and consumer and citizen-centric services in every vertical, be it agriculture, healthcare, education, *etc.*

   The pillar of demand presents an unprecedented opportunity for 'Make in India' in every industry. The number of challenges launched by various ministries such as MeitY, AIM (Atal New India Challenges), DST (Kawach) and DBT BIRAC (biotech challenges) can be fully tapped.
4. **Pillar of technology:** The best multinationals of the world are leveraging Indian talent and rushing to set up large R&D hubs in India. Aatma Nirbhar Bharat has now turned the gaze of this world class innovative talent inwards, to create products and services for the Indian market at par with developed countries.

   The use of 4.0 technologies, including 3D printing, IOT, AR/VR, biotech, cognitive computing, AI/Blockchain, *etc.* by the startup ecosystem can result in efficient products and services.
5. **Pillar of socio-economic growth:** With 22 percent of its population still below the poverty line and 44 percent of its economy still agri-based, India needs to ensure that rapid economic progress encompasses societal progress. There is need for more inclusive and responsible innovations to reach all segments of the population.

NITI Aayog is leading from the front to promote a culture of innovation and entrepreneurship in the country. Towards this end, AIM has taken a holistic approach to ensure creation of a problem-solving innovative mindset in schools and creating an ecosystem of entrepreneurship in universities, research institutions, private and MSME sector.

Initiatives of AIM

The various initiatives of AIM are:

1. **Atal Tinkering Labs** (School Level)- AIM has launched the establishment of thousands of Atal Tinkering Labs enabling students

from grade 6 to grade 12 to have access to and tinker with innovative tools and technologies like 3D printers, robotics, miniaturised electronics do it yourself kits, *etc.*

2. **Atal Incubators** (Universities, Institutions, Industry Level)- To promote creation of a supporting ecosystem for start-ups and entrepreneurs, AIM has been establishing world class incubators called Atal Incubation Centres (AICs) in universities. Institutions, corporate, *etc.* that would foster world class innovative start-ups.
3. **Atal Community Innovation Centres** (ACIC) (for serving unserved and under-served regions of India) - To promote the benefits of technology led innovation to the unreached regions of India including Tier 2, Tier 3 cities, aspirational districts, tribal, hilly and coastal areas, AIM is setting up Atal Community Innovation Centres with a unique partnership driven model wherein AIM would grant upto Rs 2.5 crores to an ACIC subject to a partner proving equal or greater matching funding.
4. **Atal New India Challenges** (Product and Service Innovations with National Impact) To create product and service innovations having national socio-economic impact, AIM has launched over 24 Atal New India Challenges in partnership with five different ministries and departments of central government.
5. **Applied Research and Innovation for Small Enterprises (ARISE)-** To stimulate MSME industry innovation to promote innovation in a phased manner in the MSME/Startup sector. AIM will be launching ARISE along with partner Ministries so that great research ideas are converted to viable innovative prototypes followed by product development and commercial deployment.
6. **Mentor of Change** (Mentorship and Partnerships with Public, Private sector, NGOs, Academia, Institutions) To enable all the initiatives to succeed, AIM has launched one of the largest mentor engagement and management program "Mentor India- The mentors of Change".

3.9. India Innovation Index (III)

Driven by 'Aatmanirbhar Bharat' and 'Make in India' initiatives, there is a strong push to empower the local manufacturing sectors that would, in turn, contribute to the production of innovative products at economical rates. Along with improving the scale of production, India has also aimed at enhancing its research capacities by introducing the new Science, Technology, and Innovation Policy 2020. These measures play a crucial role in enabling the innovative capacity of the country.

India enjoys a crucial competitive advantage in the world, with its demographic dividend. Theoretically, this demographic dividend occurs when the proportion of the working population within the total population is high, which implies that more people have the potential to be productive and contribute to the economic growth of the country.

The India Innovation Index is the outcome of extensive research and analysis for evaluating India's innovation readiness at a sub-national level, highlighting the obstacles that deter governments, businesses, and individuals from fully capturing the benefits of innovation

NITI Aayog, along with the Institute for Competitiveness released the second edition of the India Innovation Index during January, 2021, which examines the innovation capabilities and performance of the states and union territories. The first edition of the index was launched in October 2019.

The objective of the India Innovation Index is to scrutinize the innovation capacities and performance of Indian states. The index provides an extensive framework for the constant evaluation of the innovation ecosystem of the 28 Indian states and 9 union territories. The index intends to accomplish the following three functions:

1. Rank all states and union territories based on their index score
2. Identify opportunities and challenges
3. Assist in modifying governmental policies to foster innovation

The details of the India Innovation Index are provided in **Annexure-3**

3.10. Global Innovation Index (GII)

The Global Innovation Index (GII) is a joint collaborative effort between the WIPO, Cornell University, and INSEAD that annually ranks the innovation performance of more than 130 economies. As of 2020, India's rank as the 48th innovative nation is a testament to the improvements made in its innovation ecosystem.

The Global Innovation Index 2021 captures the innovation ecosystem performance of 132 economies and tracks the most recent global innovation trends.

Only Switzerland and Sweden have remained in the top three of the innovation ranking for more than a decade.

India, Kenya, the Republic of Moldova, and Viet Nam hold the record for overperforming on innovation relative to their level of development for the 11th year in a row

India (46th in 2021) moves further ahead, by two spots (48th in GII 2020), after making it into the top 50 last year. It takes second place in the lower middle-income group. India held the third position in its income group in 2019 and 2020 having entered the top three in 2019. India has also been portrayed as successful in developing sophisticated services that are technologically dynamic and can be traded internationally.

The details of the indicators of GII are given in **Annexure-4**

3.11. Other Innovation Initiatives

IPR initiatives of Ministry of Electronics and IT

Ministry of Electronics and Information Technology (MeitY) has acknowledged R&D and promotion of innovation as an integral part of the ICT ecosystem. Towards this, it has been supporting the entire value chain of R&D activities in the country ranging from the basic components to sophisticated product development in the sphere of ICT.

Recognizing the need for the protection of Intellectual Property (IP) as a vital component of innovation and scientific advance and that many of the benefits of inventions will be lost if the resulting IP is not protected, MeitY has made significant strides in recent years in creating a conducive ICT-IPR ecosystem for creation, protection, awareness and commercialization of IP as well as IP Rights.

Lighthouse India

It is an initiative of the World Bank Group to support systematic efforts to create and curate knowledge and know-how generated within India and disseminate it across the country as well as to the outside world.

Support for a Lighthouse India includes amplifying the country's role as a leader on key global development issues and leveraging its vast experience to provide for two-way learning. The World Bank Group also supports sustained dialogue among policy influencers and the private sector in the South Asia region to sharpen awareness and strengthen coalitions for action on common development challenges within and across countries.

3.12. Organisations Related to Innovations in Agriculture

A. Global Level

Agricultural Science and Technology Institution (ASTI)

In 2017, the International Food Policy Research Institute (IFPRI) and the Asia-Specific Association of Agricultural Research Institutions (APAARI) established ASTI Indo-Pacific for the purpose of strengthening ASTI's data collection, analyses and outreach in the Indo-Pacific region.

Decision makers depend on policy-relevant data and analyses to make well-informed investment decisions and to formulate high-impact policies that support the attainment of agricultural, economic, welfare, and sustainability goals. ASTI plays a critical role in this regard. Over time, ASTI has evolved from an ad hoc data-collection initiative to become a globally recognized and trusted source of policy-relevant data and analytics on agricultural research investments, spending, and capacity in low- and middle-income countries. In efforts to secure its long-term sustainability and relevance, ASTI is embarking on a new operating model that decentralizes data collection and management, targets demand-driven analyses, and draws on a multistakeholder platform of collaborators.

The ASTI Network comprise the following:

1. **Regional networks of in-country partners** responsible for the data collection, analyses, and outreach in their regions
2. **Technical and policy partners** that facilitate the integration of ASTI's outputs in relevant policy and decision making forums
3. **An ASTI Secretariat** that provides technical assistance and quality assurance, coordinates the global research agenda, and strengthens national and regional capacity in close collaboration with network partners

Tropical Agricultural Platform (TAP)

FAO is involved in establishing TAP, which is a G20 initiative to promote agricultural innovation in the tropics. To feed a growing population and meet the challenge of climate change, agricultural production must intensify in a sustainable way, and innovation is key to make that happen.

In 2012, the TAP was launched at first G20 MACS (Meetings of Agricultural Chief Scientists of G 20 States) in Mexico to address the capacity gaps for agricultural innovation in low and lower-middle income countries, in the tropics. TAP functions as a multilateral facilitation mechanism to promote greater coherence and impact of Capacity Development (CD) for Agricultural Innovation System.

TAP focuses on the development of national capacities for agricultural innovation in the tropics, where most of the developing countries are located and the capacity gap is especially wide. By helping to bridge the capacity gap, TAP aims to pave the way for agricultural innovations that meet the demands of its principal users - small farmers, small and medium-sized agribusinesses and consumers.

The Common Framework developed by the TAP proposes a practical approach to Capacity Development for Agricultural Innovation that aims at harmonizing, through an Agricultural Innovation Systems (AIS) perspective, the diversity of existing strategies. The Framework provides concepts, principles, methodologies and tools to better understand the architecture of AIS, to assess CD needs and to plan, implement, monitor and evaluate CD interventions.

Global Forum on Agricultural Research and Innovation (GFAR)

The Global Forum on Agricultural Research and Innovation (GFAR) is an open, collective movement shaping a new future for agriculture and food. GFAR makes agri-food research and innovation systems more effective, responsive and equitable, towards achieving Sustainable Development Goals. GFAR offers a forum to develop strategic partnerships and alliances for agricultural innovation among their organisations as well as with other actors in agri-food research and innovation.

Partners in GFAR include farmers and agricultural workers, civil society and NGOs, consumers, education, finance and investment institutions, UN Agencies, public sector national and international research, advisory services, private sector companies, women's groups and youth organizations. Partners engage in GFAR both directly and through their representative networks and fora.

Its governance includes representatives of international organisations like FAO, CGIAR, regional for a of national research organisations like APAARI, and of international networks and grassroots organisations representing other key stakeholders, international farmers' organisations, donor organisations, NGOs/CSOs, private sector, and advisory services.

Some of the initiatives under GFAR include:

1. **Gender in Agriculture Partnership (GAP) -** Launched at the first Global Conference on Women in Agriculture (GCWA, New Delhi, March 2012), GAP is a multi-stakeholder network for action, developed through the passion and commitment of individuals across many civil, private, governmental and inter-governmental institutions at global, regional, national and local levels working together to promote gender equity and empower women in agriculture.
2. **Global Open Data for Agriculture and Nutrition (GODAN) -** GODAN is a rapidly growing network of over 1000 global innovators and change makers across national governments, non-governmental organisations, and international and private sector organisations. A facilitator for pioneering ideas and cross-continental growth, GODAN aims to impact the lives of at least 12 million smallholders internationally by implementing innovations that help achieve the UN Sustainable Development Goal 2 - ending global hunger, achieving food security, improving nutrition and sustainable agriculture by 2030.

   GODAN supports the proactive sharing of open data to make information about agriculture and nutrition available, accessible and usable to deal with the urgent challenge of ensuring world food security. GODAN promotes collaboration to harness the growing volume of data generated by new technologies to solve long-standing problems and to benefit farmers and the health of consumers.
3. **YPARD-GFAR thematic collection action-** In 2006, Young Professionals for Agricultural Development (YPARD) was launched

to give a voice to youth in agriculture and help attract and keep them in the sector.

4. **Promoting Local Innovation (PROLINNOVA) -** Promoting Local innovation in ecologically-oriented agriculture and NRM, known as PROLINNOVA, is an NGO-initiated international learning network to promote local innovation in ecologically oriented agriculture and Natural resource management. Its HQ is Addis Ababa, Ethiopia.

The PROLINNOVA network give increased attention to identifying and documenting the local innovation processes and the innovations that result from them. This is done to increase awareness of the relevance of local innovativeness for meeting the needs of farming families and communities and to encourage development agents and scientists to interact and support this local innovativeness.

Innovation Policy Platform (IPP)

The Innovation Policy Platform (IPP), jointly developed by the OECD and the World Bank, aims to build a web-based, open-data interactive platform to facilitate collective learning processes around science, technology and innovation (STI) policy. Its goal is to provide its users with support in analysing innovation systems and policies and in shaping future policy designs. It will also facilitate collective learning processes of both the principles as well as the "how to" aspects of innovation policy, tailored to the needs of developing and developed countries.

B. National Level

The post-liberalization economy in India witnessed rapid growth, owing to the structural reforms initiated to attract foreign investment and encourage competition in India's buoyant domestic market. Spearheading this process was the National Science and Technology Entrepreneurship Development Board (NSTEDB), established in 1982, and the National Innovation Foundation (NIF) in 2000, both set up under the aegis of the Department of Science and Technology, Government of India. While NSTEDB was at the forefront of promoting science and technology parks and incubators through technology-based companies, NIF focused on strengthening the evolution, sustenance, and diffusion of grassroots innovation.

The innovation agenda has been furthered in recent times through NITI Aayog's Atal Innovation Mission, DST-NIDHI's PRAYAS (PRomoting and Accelerating Young and ASpiring technology entrepreneurs), Digital India, and Startup India. These initiatives have encouraged the entrepreneurial spirit while enabling access to the requisite resources to harness this youthful energy. The goal has been to make innovation a part of India's policy agenda, in order to develop and strengthen the existing innovation ecosystem that drives positive impact. The details are presented in the appropriate chapters.

3.13. Challenges of Indian Innovation Ecosystem

Abhyankar (2014) had identified certain key challenges faced by the Indian innovation ecosystem and entrepreneurship in general, some of which holds good in the present situation also. The challenges are briefly presented:

1. **Fragmented policy and policy implementation:** The new policy of 2020 takes care of this to a large extent.
2. **Inadequate funding for R&D:** The national funding for R&D is low compared to developed countries. However, there is significant improvement during recent years, which is a welcome step. R&D expenditure incurred by industry is also low compared to developed nations.
3. **Difficult and cumbersome procedure for funding:** Although funding is available from banks and other public sources, the procedure for accessing and availing such funds are often complex, cumbersome, lengthy, bureaucratic and not conducive to innovation and entrepreneurship development. The financial system seeks immediate returns, which are often uncertain, late or not quantifiable immediately.
4. **Angel, venture capital and seed funding:** Despite many angel networks operating in India, only limited deals are made each year. The magnitude of such deals is very low compared to the numbers from other developed countries. The early stage venture capital investment is also comparatively low. Presently, there is a slight improvement in this aspect.
5. **Weak linkages between stakeholders:** The linkages between industry, especially medium and small-scale enterprises and R&D or academic institutions are not strong. Industry requires proven technologies, but the institutions can only offer technologies at considerably earlier stages (*i.e.*, at mostly a laboratory or pilot scale), which means there is still much work to be done to bring the technologies to market. There is also considerably less funding and mentorship support available from the private sector. There is no easy exchange of manpower between the industries and academia or R&D institutions, which limits their capacity for mutual understanding and technology transfer.
6. **Non-conducive education system:** The general education system is still too focused on grades and careers and is not oriented toward innovation and entrepreneurship. This situation is further worsened by the inherent problems of lack of infrastructure and good facilities in the educational institutions; delays in the funding system; and delays in the funds or other support reaching innovation projects. A few high-end academic institutions such as Indian Institutes of Technology and similar institutions may be exceptions. However, Atal Innovation

Mission takes care of this to a great extent. Establishment of Innovation Universities is also another step in the right direction.

7. **Poor infrastructure facilities in villages:** Basic infrastructure facilities such as electricity, internet, roads and rail, and even the availability of a skilled workforce, are not evenly distributed in India and often weak in smaller cities or towns and rural parts of the country. There is less scope for innovation and entrepreneurship to flourish in such areas. Atal Community Innovation Centres are established to address this problem.
8. **Inadequate protection of intellectual property rights:** In India, the intellectual property regime is weak. Innovators do not generally seek protection for their intellectual property unless forced to. For most entrepreneurs, patents and other forms of protection take too long and cost too much. Patent literacy is very low, even among educated innovators, and there is a lack of expert help available, except in the medicine and pharmaceutical industry.

REFERENCES

Abhyankar, R. (2014) The Government of India's Role in Promoting Innovation through Policy Initiatives for Entrepreneurship Development. *Technology Innovation Management Review*, 4(8): 11-17. http://doi.org/10.22215/timreview/818

FAO (2018) FAO's work on Agricultural Innovation. http://www.fao.org/3/CA2460EN/ca2460en.pdf

Granstrand, Ove and Holgersson, Macus (2020) Innovation ecosystems: A conceptual review and a new definition, Technovation, Elsevier, vol. 90

Global Forum for Agricultural Research and Innovation (2018) An independent forward-looking learning review, Final draft report, October 2018. https://www.researchgate.net › publication › 334129266_.

Jackson, J Deborah (2013) What is an Innovation Ecosystem? https://www.semanticscholar.org>paper>what-is-an-innovation-ecosystem

Institute for Competitiveness (IFC) and NITI Aayog (2020) India Innovation Index 2020, NITI Aayog, New Delhi

Kimmo Halme (2014) How to build an innovation ecosystem? Some lessons from Finland, World Bank Seminar on 'Drivers of Private Sector Innovation' Istanbul, May 26, 2014

Lundvall, B. A. (1992). National systems of innovation: Towards a theory of innovation and interactive learning. Pinter Publishers, London.

Michael E Porter (1980) Competitive Strategy: Techniques for Analyzing Industries and Competitors, Simon and Schuster

Moore, J F (1993) Predators and prey: A new Ecology of competition, Harvard Business Review, 71: 75-86

Renate Kloeppinger-Todd, Manohar Sharma (2010) Innovations in rural and agricultural finance, International Food Policy Research Institute, Washington D C. http://dx.doi.org/10.2499/0896296687

Roseboom, J. (2011) Supranational collaboration in agricultural research in sub-Saharan Africa, Paper ASTI/IFPRI-FARA Conference, Ghana, December 2011.

Yehia Zahran, Hazem S. Kassem, Shimaa M. Naba and Bader Alhafi Alotaibi (2020) Shifting from Fragmentation to Integration: A Proposed Framework for Strengthening Agricultural Knowledge and Innovation System in Egypt *Sustainability* **2020**, *12*, 5131; doi:10.3390/su12125131

WIPO Global Innovation Index (GII). https://www.wipo.int › global_innovation_index

Chapter 4

Diffusion of Technological Innovations in Agriculture

Everett M Rogers had substantially contributed to diffusion research through his two main publications, 'Diffusion of Innovations' (Rogers, 1962) and 'Communication of Innovations' (Rogers and Shoemaker, 1971), considered the Bible of diffusion studies. Rogers' model of innovation studies evolved in a particular institutional context, based on well-known case studies, notably the adoption and diffusion of hybrid corn in the USA in the 1950's. These case studies simplified a complex process of diffusion into a step-wise, linear sequence of activities described without reference to the way agendas of different stakeholders were addressed, and how technology priorities were selected and promoted (Sulaiman and Hall, 2002). Innovation was understood only as a new technology at that time, whereas we now have a better understanding of the term 'innovation', encompassing as it does the factors affecting demand for, and use of, knowledge in novel and useful ways (World Bank, 2006).

4.1. Meaning of Diffusion

Technological innovations have greatly shaped agriculture throughout time. The change in agriculture started in India with the Green revolution, trailed by accomplishments of large achievements such as white revolution, blue revolution, yellow revolution, pink revolution, *etc.*, all adopting scientific technological innovations.

Diffusion of innovations is a concept that seeks to explain how, why and at what rate new ideas and technologies spread in society. It is a social process that occurs among people in response to learning about an innovation such as a new evidence-based approach for improving farming. Diffusion is the process by which an innovation is communicated through certain channels over time among the members of a social system (Rogers, 2003).

An innovation is an idea, practice, or object perceived as new by an individual or other unit of adoption. The diffusion process typically involves both mass media and interpersonal communication channels. In today's world, information technologies such as the internet and cell phones, which combine aspects of mass media and interpersonal channels, represent formidable tools of diffusion. Artificial intelligence based technologies have further hastened the diffusion process.

In the 1950s, innovation in agriculture was essentially thought of as a phenomenon of adoption and adaptation. Science was perceived as external to the socio-economic system, independent and neutral, and a source of innovation, whereas traditional knowledge was seen as a barrier to the spread of agricultural progress. Farmers were urged to adopt technologies communicated to them by the development agencies.

Subsequently, it was felt that technologies should not be thrust on the farmers (beneficiaries), instead, their interests and their environment had to be taken more into account, and thus it became necessary to modify the methods of intervention. With more encompassing approaches being required, the discourse among researchers and development agencies gave rise to two new frameworks of thought: AKIS (Agricultural Knowledge and Information Systems) and AIS (Agricultural Innovation Systems) (Klerkx *et al.*, 2012). In both these frameworks, the interactive innovation model emerged as the solution in contrast to the linear model.

The AKIS frameworks focuses on the exchange of knowledge and information to sustain the innovation process. The AIS approach is intended to be even more inclusive; it takes into account all the actors who participate, directly or indirectly, in innovation processes (input suppliers, actors of supply chains, banks, policymakers, *etc.*).

4.2. Concept of Innovation Diffusion

The concept of Diffusion of Innovation (DOI), developed by E.M. Rogers in 1962, is one of the oldest social science concepts. It originated in the discipline of communication to explain how, over time, an idea or product gains momentum and diffuses or spreads through a specific population or social system. The end result of this diffusion is that people, as part of a social system, adopt a new idea, behaviour, or product. The key to adoption is that the person must perceive the idea, behaviour, or product as new or innovative. It is through the perception of 'newness' that diffusion is possible.

Adoption of a new idea, behaviour, or product (*i.e.*, innovation) does not happen simultaneously in a social system; rather it is a process whereby some people are more inclined to adopt the innovation than others. Researchers have found that people who adopt an innovation early have different characteristics than people who adopt an innovation later. When promoting an innovation to a target population, it is important to understand the characteristics of the target population that will help or hinder adoption of the innovation.

Diffusion of innovation theory is built on the premise that any commercial consumer marketplace has different types of customers, who vary on their enthusiasm for a particular product, and for trying out that product. The theory also presupposes that so-called innovative products diffuse into the marketplace not on a straight path, but in wave after wave of consumer acceptance, starting with innovators, then moving on to early adopters, early majority, late majority, and laggards at the end of the line.

Several contextual aspects of diffusion typically go unstudied. Competing or complementary innovations are important, since potential adopters usually have a choice in what to adopt. Failures are important, since most innovations do not diffuse. Deceleration is important in two ways, since the decision to adopt an innovation often means abandoning a prior one, and nonadopters have their decision to reject an innovation socially confirmed.

Diffusion or the lack thereof is often well explained by three general sets of variables: each innovation's set of pros and cons, or attributes; the characteristics of adopters, especially potential adopters' perceptions of opinion leaders' reactions, or social influence; and the larger social and political context, including the salience of issues related to the innovation, how proponents and opponents frame the meaning of the innovation, and the timing of its introduction.

4.3. Adopter Categories

There are **five established adopter categories**, as given by Rogers. The innovation adoption follows a bell-shaped curve, which can be divided into five different categories. Though the majority of the general population tends to fall in the middle categories, it is still necessary to understand the characteristics of the target population. When promoting an innovation, there are different strategies used to appeal to the different adopter categories.

1. **Innovators:** These are people who want to be the first to try the innovation. They are venturesome and interested in new ideas. These people are very much willing to take risks, and are often the first to develop new ideas. Mostly, nothing needs to be done to appeal to this population.
2. **Early Adopters:** These are people who represent opinion leaders. They enjoy leadership roles, and embrace change opportunities. They are already aware of the need to change and so are very comfortable

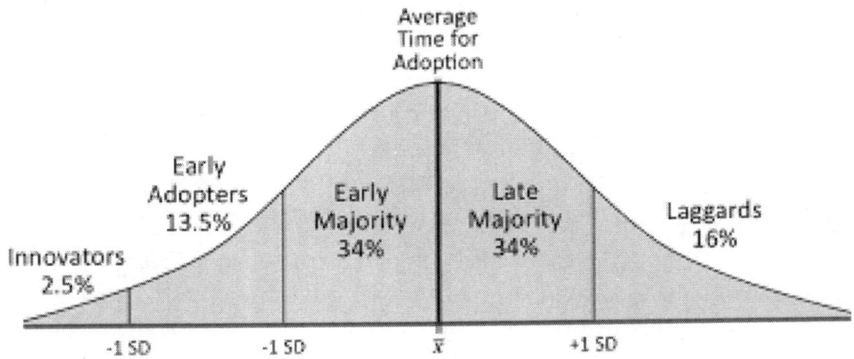

Figure 2: Bell Shaped Curve.

adopting new ideas. Strategies to appeal to this population include how-to manuals and information sheets on implementation. However, they do not need information to convince them to change.

3. **Early Majority:** These people are rarely leaders, but they do adopt new ideas before the average person. They typically need to see evidence and get convinced that the innovation works before they are willing to adopt it. Strategies to appeal to this population include success stories and evidence of the innovation's effectiveness.
4. **Late Majority:** These people do not favour change, and will only adopt an innovation after it has been tried by the majority. Strategies to appeal to this population include information on how many other people have tried the innovation and have adopted it successfully.
5. **Laggards:** These people are bound by tradition and very conservative. They are very skeptical of change and are the hardest group to bring

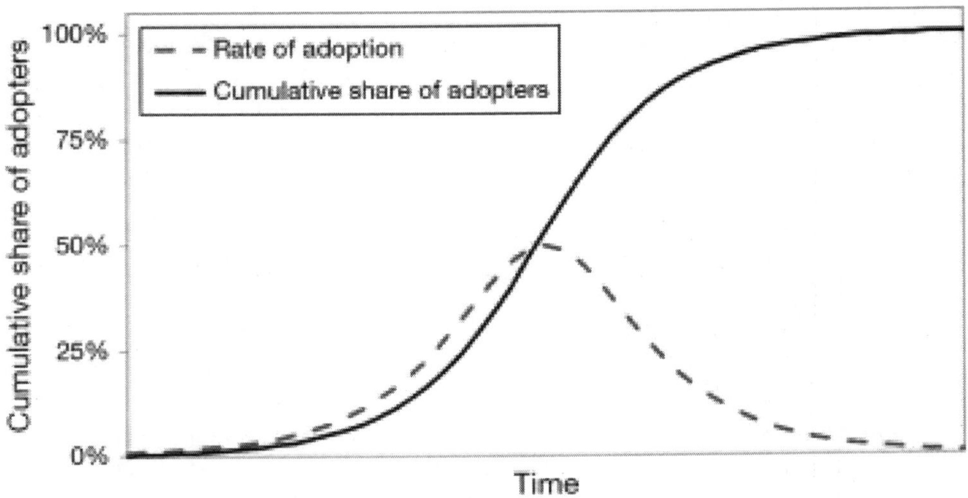

Figure 3: S Shaped Curve.

on board. Strategies to appeal to this population include statistics, fear appeals, and pressure from people in the other adopter groups.

Rogers holds that the technologies tend to follow an 'S' shaped adoption curve. Adoption starts slowly, goes through an accelerated and scaled interval, before it stabilizes and matures.

Box 4.1

In seeking to understand processes of agrarian development, the notion of a treadmill was first introduced by Willard Cochrane, an American agricultural economist, in 1958. He became Chief Economist in the U S Department of Agriculture during the Kennedy Administration and played a distinctive role within American agricultural economics. His theory, though first introduced in the I950s, is best encapsulated in his 1979 account of 'The Development of American Agriculture'. Drawing on findings from rural sociological research into the diffusion and adoption of agricultural innovations, Cochrane argued in his 'theory of the agricultural treadmill' that non-risk-averse early adopters, who were the first farmers to adopt new technologies, benefitted from lower unit costs of production associated with an increase in output, and these lower costs raised net returns.

When it comes to adoption of a new technology, Cochrane (1979) divided the farming population into three subgroups- early adopters, followers, and laggards. The early adopters may be a small fraction of the population, in which case the impact of their adoption decision on aggregate supply and, thus, output prices is relatively small. Therefore, these individuals stand to profit from the innovation.

The followers are the large share of the farm sector who tend to adopt during the take-off stage of the innovation. Their adoption choice will eventually tend to reduce prices, which reduces profits as well. This group of adopters may gain or lose as a result of innovation.

Finally, the third group of farmers (laggards) are those who either adopt at the lag stage of the adoption process or do not adopt at all. These individuals may lose from technological change. If they do not adopt, they reduce the same quantity as before, at low prices; and if they adopt, the significant price effect may sweep the gain associated with higher yields.

Thus, Cochrane argues that farmers, on the whole, are not likely to gain from the introduction of innovation in agriculture, except for a small group of early adopters. Introduction of new technology may lead to structural change and worsen the lot of some of the small farms. The real gainers from technological change and innovation in agriculture are likely to be consumers, who pay less for their food bill.

Source: Levins, R.A. and Cochrane, W.W. (1996). The Treadmill Revisited, Land Economics, 72(4): 550-553.

4.4. Diffusion of Innovation Process

Rogers described the innovation-decision process as 'an information-seeking and information processing activity, where an individual is motivated to reduce uncertainty about the advantages and disadvantages of an innovation'. According to him, the innovation-decision process involves five steps: (1) knowledge, (2) persuasion, (3) decision, (4) implementation, and (5) confirmation. These stages typically follow each other in a time-ordered manner.

1. **Knowledge Stage**: The innovation-decision process starts with the knowledge stage. In this stage, an individual learns about the existence of innovation and seeks information about the innovation. 'What?,' 'how?,' and 'why?' are the critical questions in the knowledge phase. During this phase, the individual attempts to determine 'what the innovation is and how and why it works'. Three types of knowledge are: (1) awareness-knowledge, (2) how-to-knowledge, and (3) principles-knowledge. At this stage, the marketers will be trying to enhance awareness of the product and provide more information so that the prospective adopter moves to the second stage.
2. **Persuasion Stage**: The persuasion stage occurs when the individual develops a negative or positive attitude toward the innovation. The individual shapes attitude after knowing about the innovation and understands more about the same. It is the point at which the prospective adopter is open to the idea of purchase, who is actively seeking more information which will influence his/her eventual decision. This is the stage at which marketers will be trying to convey the benefits of the product in much detail.
3. **Decision Stage**: The individual makes a decision after weighing up the pros and cons of the innovation. Rogers cites this as the most difficult phase on which to acquire intelligence. People do not make rational decisions in many instances. They make decisions based on their underlying perceptions and feelings and after the decision, they attempt to rationalize that decision. At the decision stage in the innovation-decision process, the individual chooses to either adopt or reject the innovation. While adoption refers to full use of an innovation as the best course of action available, rejection means not to adopt an innovation.
4. **Implementation Stage:** At the implementation stage, an innovation is put into practice. However, an innovation brings the newness in which some degree of uncertainty is involved in diffusion. Uncertainty about the outcomes of the innovation still can be a problem at this stage. Thus, the implementer may need technical assistance from change agents and others to reduce the degree of uncertainty about the consequences.

This stage is when the adopter is clear as to whether or not the product is actually useful. He/she may also seek out further information to either support the use of the product or to better understand the product in context. The quality of the implementation experience is going to be determined, to a lesser or greater extent, by the ease of access to information and the quality of that information.

5. **Confirmation Stage**: The innovation-decision already has been made, but at the confirmation stage the individual looks for support for decision. This decision can be reversed if the individual is exposed to conflicting messages about the innovation. However, the individual tends to stay away from these messages and seeks supportive messages that confirm the decision. Thus, attitudes become more crucial at the confirmation stage.

   This phase will normally involve a personal examination of the product and also a social one (the user will seek confirmation from their peers, colleagues, friends, *etc.*)

4.5. Process of Adoption

Ryan and Gross (1943) were the first to recognize that the adoption of new idea consisted of stages. They distinguished between awareness of hybrid seed corn, conviction of its usefulness, trial acceptance and complete adoption of the innovation. Wilkening (1953) identified four adoption stages- awareness, obtaining information, conviction and trial, and adoption. Johnson and Haver (1955) identified the steps of decision-making as: (i) Observing the problem (ii) Making analysis of it (iii) Deciding the available course of action (iv) Taking one course (v) Accepting the consequences of the decision.

The North Central Rural Sociology Subcommittee for the study of diffusion of farm practices (1955) identified five stages of the adoption process, which received world-wide attention. These are: (i) Awareness (ii) Interest (iii) Evaluation (iv) Trial (v) Adoption. The stages of adoption are dynamic and not static. The same five stages do not occur with all the adopters and all the practices. Sequence is not always the same. Sometimes one stage appears more than once. In some cases, some stages are too short as to be imperceptible, and in other cases some stages seem to be skipped. There are no clear-cut differences and sometimes the whole process is capsuled and look like a unit act.

Adoption process focuses on the mental processes through which an individual passes from first hearing about an innovation to its final adoption. The different stages are:

1. **Awareness:** The individual becomes aware of the new product, but lacks complete information about it. Awareness leads to interest and the individual seeks more information about the new product.

Box 4.2

Ryan and Gross (1950) conducted an analysis of the conditions and processes under which important technical innovation, hybrid corn, was adopted in two prosperous agricultural communities in Iowa, USA. Three general problems were studied: (1) the time pattern by which use of the seed spread; (2) the functions and importance of its diffusion agencies or media by which the seed spread; and (3) the relationship of characteristics (personal, economic and social) of farm operators to the rapidity with which they adopted the new seed. The salient results are:

1. The diffusion pattern was made up of three periods: (a) a long period of slow initial growth, (b) rapid rise in adoption and (c) a brief decline as the most resistant adopters accepted the technique.
2. The earliest adopters performed a special function for their communities in their roles as "experimenters."
3. Most adopters first accepted the seed in small quantities, gradually increasing their use as the years went by. Even late adopters usually went through such a period of trial use in spite of the currently larger plantings by those who had tried the seed earlier.
4. The diffusion media by which farm operators first learned of hybrid corn tended to differ from the media most influential in leading them to use it. This suggests a functional classification of diffusion agencies, as introductory and as activating media.
5. The major single source of original knowledge about hybrid seed was salesmen. The most influential single source of knowledge was "neighbours.".
6. Size of farm enterprise, education, youthfulness and amount of social participation of the operator were all associated with early adoption.
7. The forms of social participation associated with early adoption indicate a greater susceptibility to technological innovation among farm operators who had a more secular pattern of social life.
8. The analysis suggests that those members of an agricultural community most responsive to secularized contacts may represent a social type also most responsive to innovation in agricultural practice.

Source: Ryan, Bryce and Gross, Neal (1950) Acceptance and diffusion of hybrid corn seed in two Iowa communities, Research Bulletin (Iowa Agriculture and Home Economics Experiment Station): Vol. 29: No. 372, Article 1. Available at: http://lib.dr.iastate.edu/researchbulletin/vol29/iss372/1.

2. **Interest:** The individual seeks more information about the product and gets motivated to obtain further information about the features, utility, *etc.* of the product
3. **Evaluation:** During evaluation, the individual considers whether the product will satisfy certain criteria for meeting the specific needs and decide whether to try it or not.
4. **Trial:** The individual tries the new product on a small scale. During this stage, potential users determine the usefulness of the product under various conditions.
5. **Adoption:** The individual decides to make full and regular use of the new product. The continued use of the product is referred to as adoption.

4.6. Diffusion and Adoption

Diffusion is a macro process related to the spread of an innovation from its source to the final consumers, whereas adoption is a micro process focusing on the stages through which an individual passes in deciding to accept or reject an innovation.

While adoption is the process by which a user begins and continues to use a product, diffusion is considered a measure of the rate of adoption. Diffusion considers the relationship not just between any given user and a product but the relationship between all users, each other and the product.

The original diffusion research was done as early as 1903 by the French sociologist Gabriel Tarde who plotted the original S-shaped diffusion curve and by German and Austrian anthropologists Friedrich Ratzel and Leo Frobenius. Theories of innovation diffusion have been used to increase the adoption of innovative products and practices by a number of professionals. These professionals belong mainly to the disciplines of agriculture, sociology and marketing.

There are many theories of diffusion applicable in the fields of agriculture, sociology and marketing. Although every theory has a distinct rationale behind it, the universality of every theory is challengeable except the rate of adoption theory. This theory states that innovations are diffused over time in a pattern that resembles an S-shaped curve that is universal.

4.7. Theories of Innovation Diffusion

Laws of Imitation

The early theory of adoption probably is by Gabriel Tarde, considered as one of the forefathers of sociology and social psychology. Tarde observed certain generalizations about the diffusion of innovations, which he termed "the laws of imitation". According to Laws of Imitation,

1. The more similar an innovation is to those ideas that have already been accepted, the more likely the innovation is to be adopted
2. Rate of adoption of a new idea usually followed an S-shaped curve over time. At first, only a few individuals adopt a new idea, subsequently the rate of adoption spurts as a large number of individuals accept the innovation, and finally the adoption rate slackens.
3. An innovation is first adopted by an individual who is socially close to the source of the new idea, and then it spreads gradually from higher-status to lower-status individuals.

Rogers (1983) remarked that Tarde's (1903) S-shaped curve is important because "most innovations have an S-shaped rate of adoption". The variance lies in the slope of the "S". Some innovations diffuse rapidly creating a steep S-curve; other innovations have a slower rate of adoption, creating a more gradual slope of the S-curve.

The British and German-Austrian Diffusionists' Theory

A group of early anthropologists that evolved in England and in Germany-Austria soon after the time of Gabriel Tarde took up further research in diffusion. These anthropologists are called the "British diffusionists" and the "German-Austrian diffusionists." The viewpoint of each group was similar. Diffusionism was the point of view in anthropology that explained change in a given society as a result of the introduction of innovations from another society. The diffusionists claimed that all innovations spread from one original source, which, of course, argued against the existence of parallel invention. They proposed that all social change could be explained by diffusion alone. The dominant viewpoint is that social change is caused by both invention and diffusion, which usually occur sequentially (Rogers, 1983).

Individual Innovativeness Theory

The individuals are classified into different categories according to their innovativeness as: innovators, early adopters, early majority, late majority, and laggards. Innovativeness is the degree to which an individual or other unit of adoption is relatively earlier in adopting new ideas than other members of a system (Rogers, 1983). In each adopter category, individuals behave in a similar way in terms of their innovativeness. While this is a widely accepted theory, perhaps it is the most criticized one also.

Considering the consistency in innovativeness of an individual towards all the innovation, Rogers himself said that there is no clear-cut evidence as to whether or not innovating behaviour is completely consistent. However, he postulated that it is doubtful whether an individual who is innovative in one aspect is laggard for another idea. Further, the adopter categories also change over time as adopter categorization is similar to a snap-shot that pictures an individual at one time. He need not necessarily remain in the same position in a social structure at later point in time. So there is no general innovator or universal laggard for all innovations and over period of time as is often implicitly assumed by some diffusion researchers. (Chamala *et al*,1980).

Based upon their research on different innovations, Siddaramaiah and Nithyashree (2005) presented new adopter categories and nomenclature. They divided the adopters into four groups as: (i) Pioneers (ii) Rationalists (iii) Imitators (iv) Murmurers. As this model has identified two groups of adopter categories on either side of the mean, the distribution is said to be in equilibrium. Hence, the model is called as PRIM (E) model of adopter categories in which E means equilibrium. The first group, Pioneers consists of 15 per cent of adopters. They are venturesome and originators in testing and adopting any new idea they come across. The second group rationalists constitute a highest of 35 per cent is sensible and try to explain everything by reasoning. The imitators and murmurers are 32 and 18 per cent respectively. There is nothing new and significant in this model except the merging of first two categories given by Rogers *i.e.* innovators and early adopters.

Rate of Adoption Theory

A widely-used diffusion theory discussed by Rogers (1983) is the theory of Rate of Adoption. Rate of Adoption theory states that innovations are diffused over time in a pattern that resembles an S-shaped curve. Rate of Adoption theorizes that an innovation goes through a period of slow, gradual growth before experiencing a period of relatively dramatic and rapid growth. The theory also states that following the period of rapid growth, the innovation's rate of adoption will gradually stabilize and eventually decline. He defined the rate of adoption as the relative speed with which an innovation is adopted by members of a social system.

Theory of Cultural Lag

Cultural lag was coined by sociologist William F. Ogburn, which refers to the notion that culture takes time to catch up with technological innovations, and that social problems and conflicts are caused by this lag. According to Ogburn, cultural lag is a common societal phenomenon due to the tendency of material culture to evolve and change rapidly and voluminously while non-material culture tends to resist change and remain rigid for a far longer period of time. Due to the opposing nature of these two aspects of culture, adaptation of new technology becomes rather difficult. Cultural lag theory suggests that a period of maladjustment occurs when the non-material culture is struggling to adapt to new material conditions. The theory resonates with the ideas of Technological Determinism, in that it assumes that technology has independent effects on society at large. Ogburn posited four stages of technical development: invention, accumulation, diffusion, and adjustment.

1. Invention is the process by which new forms of technology are created.
2. Accumulation is the growth of technology.
3. Diffusion is the spread of an idea from one cultural group to another, or from one field of activity to another.
4. Adjustment is the process by which the non-technical aspects of a culture respond to invention.

Cultural lag creates problems for a society in a multitude of ways. The issue of cultural lag tends to permeate any discussion in which the implementation of some new technology is a topic. For example, the advent of bioengineering research has given rise to many new, potentially beneficial medical technologies; however these new technologies have also raised serious ethical questions about the use of bioengineering in medicine.

Bass Theory

A theory of diffusion was given by Bass (1969), the most popular model in the field of marketing, which is a mathematical model of diffusion. It consists of a simple differential equation that describes the process of how new products get adopted in a population. The model presents a rationale of how current adopters and potential adopters of a new product interact.

Bass model assumes that potential adopters of an innovation are influenced by two means of communication-mass media and word of mouth. In its development, it further assumes that the adopters of an innovation comprise two groups. One group is influenced only by the mass-media communication (external influence) and the other group is influenced only by the word-of-mouth communication (internal influence). Bass termed the first group "Innovators" and the second group "Imitators."

There are certain assumptions of this model as:

- ✰ Diffusion process is binary (consumer either adopts, or waits to adopt)
- ✰ Constant maximum potential number of buyers is given as N.
- ✰ Eventually, all N will buy the product
- ✰ No repeat purchase, or replacement purchase
- ✰ The impact of the word-of-mouth is independent of adoption time
- ✰ Innovation is considered independent of substitutes
- ✰ The marketing strategies supporting the innovation are not explicitly included

The Bass model proposes that the likelihood of purchase of a product at time 't' denoted by L (t) is a linear function:

$$L(t) = p + \frac{q}{N} N(t)$$

where,

p = Coefficient of innovation (or coefficient of external influence)

q = Coefficient of imitation (or coefficient of internal influence)

N(t) = Total number of adopters of the product up to time t

N = Total number of potential buyers of the new product

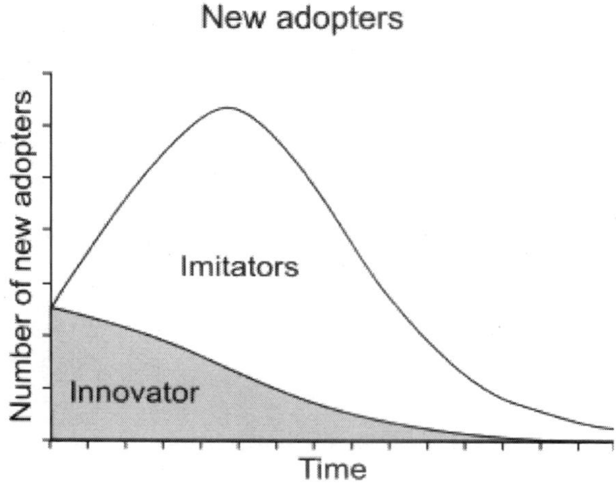

Figure 4: Bass Model.

Model of Eastlick and Lotz

Gatignon and Robertson's model (1985) was developed to explain the process of diffusion of innovations, which assumes that the variables related to the intrinsic characteristics of each individual influence the process of adopting innovation. Eastlick and Lotz (1999) have given an improved version of that model. It was designed to investigate the personal characteristics, attitudes and habits of innovators facing adoption of new technologies.

The model of Eastlick and Lotz shows that the elements, which have a direct influence on the decision of adopting or non-adopting a new technique or technology, are:

1. Attitude towards innovation. This attitude can be positive or negative depending on an assessment of innovation attributes made by the individual in question.
2. Habits which, up till now, involve some usage behaviour and which should change as consequence of adopting a new system.

The theory states that the more the new behaviour is close to old habits, its adoption becomes easier. This model takes into account the interrelation between attitude and perceived risk. On the one hand, there are attitudes that people manage more easily felt risks than others and proceed with the adoption of an innovation. On the other hand, a high perceived risk may deter some individuals from adopting this innovation. This model also implies that attitude is formed under the influence of three factors: old habits, the innovative features of the person and the intrinsic characteristics of the innovation. (Gazbar, 2013)

Technology Acceptance Model (TAM)

The technology acceptance model (TAM) was developed by Davis (1989) to study intentions of adopting an information system. To the theory of innovation diffusion of Gatignon and Robertson, Davis has included two new factors: ease of use and perceived usefulness of the system. According to Taylor and Todd (1995), the practical usefulness of this model lies in the fact that the constructs "Perceived usefulness" and "Ease of use" are the key factors for future adoption and use and on which the system designers have some degree of control. Ajzen and Fishbein (1977) consider that intent is the only determinant of the future use of the studied technology. Davis (1989) justifies this new feature referring to workplace reality where an employee may develop a negative attitude vis-à-vis a given technology but uses it anyway because he/she believes that it is advantageous in terms of performance.

The more technology is easy to use, the more useful it will be and more positive the attitude and intention to use it later are. The main feature of this model is that it brings about a small number of factors to explain the use of a technology and which are easy to understand and handle. These factors are specific, but generalizable. (Gazbar, 2013)

Theory of Planned Behaviour (TPB)

The theory of planned behaviour (TPB) was developed by Ajzen (1991) which considers intention as a combination of three variables:

1. The attitude adopted by a person before opting for a behaviour or action.
2. The subjective norms imposed by social pressure which each individual faces in social environment and which dictate some course of action.
3. The perception of control that each individual has on the adopted system and this in terms of their own abilities and available resources.

Hype Cycle

An American research and advisory firm called Gartner inc proposed this theory which explains what happens to an innovation when it is hyped through a trigger and subsequently how it goes through various stages to full acceptance. Rather than putting the blame on potential adopters or non-adopters, this theory talks about the intrinsic value of the innovation. Gartner's hype cycle gives an insight on how an innovation evolves over time. It represents the maturity, adoption and business application of new technology. The hype cycle commences with the trigger, followed by a peak of inflated expectations, goes through a trough of disillusionment and a slope of enlightenment, and finally arrives at the plateau of productivity. The hype cycle curve will be different for different innovations. Each year the firm comes up with various hype cycles in various domains for emerging technologies. Some researchers consider it a better method for studying innovations (Gartner Research 2018).

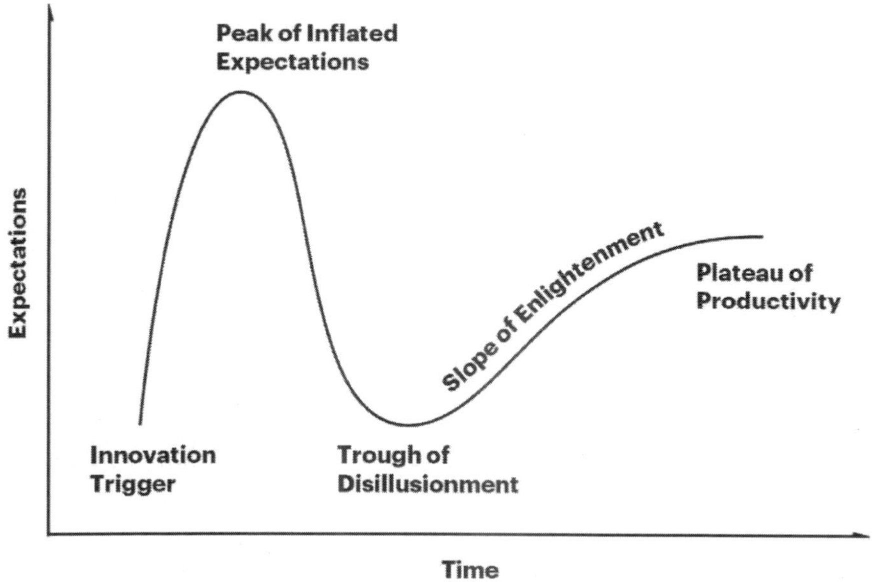

Figure 5: Hype Cycle.

Cyclic Innovation Model

Berkhout (2017) conceived this model and remarked that the successful market introduction of products and services is a cyclic process with interactions between different actors from various disciplines. This model reconciles the 'technology push' model and the 'market pull' model and explains as to why a holistic multi-disciplinary approach is needed to facilitate an effective innovation system.

Maloney's 16 per cent Rule

The 'Maloney's 16 per cent ' rule is used for accelerating the diffusion of any innovation. The rule states that once 16 per cent adoption of an innovation has occurred, one must change the media and messaging strategy from one based on scarcity to one based on social proof in order to accelerate through the chasm to the tipping point (Maloney, 2010).

Multiplicative Innovation Adoption (MIA) Model

Phipps and Simmons had proposed a model of innovation adoption process, which is based on the Rogers and Shoemaker criteria. Their idea was to understand what makes an innovative product, service or idea successful. To assess an innovation using the MIA model, it will be necessary to rate the innovation (product, service or idea) on the Rogers and Shoemaker criteria.

This is done by using the following scheme:

1. **Relative advantage:** The rating under this is referred to as RA. If the relative advantage is high, the innovation is rated 3, if it is medium, it is rated '2' and if it is low then '1'.
2. **Compatibility:** This is referred to as 'CT' rating. If the compatibility is high, a rating of '3' is given, if it is medium, a rating of '2' is given and '1' if it is low.
3. **Complex:** This is referred to as 'CL' rating. The rate of innovation is '3' if it is simple, '2' if it is of medium complexity and '1' if it is judged to be highly complex.
4. **Facilitates Trial:** This rating is referred to as TR. If the opportunity to offer trial is high the rate of innovation is '3', '2' if it is judged to be medium and ' 1' if it is low.
5. **Observability:** This rating is called 'OB'. If the observability is high, the innovation is rated '3' if it is judged to be medium it is '2' and ' 1' if it is low.

Under the MIA model, it is assumed that each of the criterion makes an equal contribution to the success or otherwise, of the innovation. Another assumption is that they combine in such a way that if there were positive ratings on more than one criterion, this will have a multiplicative effect on the success of the innovations.

4.8. Application of the Diffusion of Innovations (DOI) Approach in Extension

The DOI approach describes how and why individuals adopt or reject new things (Rogers, 2003). As communication is located at its core, the DOI is characterized as a social process. Information about a new product, practice, or idea reaches potential adopters through interpersonal and mediated communication. From there on, provided various conditions are met, the knowledge gained is translated into practice. Communication thus accompanies the innovation from its market release, through its entrance into a social system (knowledge transfer), up to its translation into something used (implementation), and diffusion in the social system (full adoption) (Valente and Rogers, 1995). Adoption by initially reluctant individuals is facilitated through interpersonal communication with its satisfied adopters.

The innovation process, with the endpoint as full adoption, is a rather lengthy one, especially for farmers. It is estimated that it takes about five years for a farmer to try out an innovation after hearing about its existence, and another 10 to 15 years until its full adoption in practice (Valente and Rogers, 1995). Such disheartening estimates fuelled research into the factors able to facilitate a more prompt adoption. Recent studies suggested a wide array of fruitful ways to employ the DOI to improve our knowledge about the factors influencing the adoption decisions by the farmers.

4.9. Attributes of Innovation

There are five attributes based on which an innovation is judged: (1) relative advantage, (2) compatibility, (3) complexity, (4) trialability, and (5) observability. Rogers (1983) stated that individuals' perceptions of these characteristics predict the rate of adoption of innovations.

The more positively people evaluate the attributes of an innovation, the more likely they are to adopt it. First, the innovation would have to be perceived as better than the product/practice it replaces (relative advantage). Second, it should appear easy to use and/or reduce the difficulties and inconveniences related to the use of the established product or practice (complexity). Third, the end users would have to be offered the opportunity to try out the innovation without being obliged to adopt it (trialability). Fourth, the innovation should have led to observable benefits for those who had already adopted it (observability). Fifth, the innovation would have to be consistent with existing values, experiences, and needs (compatibility). The attributes are elaborated below:

Relative Advantage

It is the degree to which an innovation is perceived as being better than the idea it supersedes. The cost and social status motivation aspects of innovations are elements of relative advantage. However Burkman's theory of a user-oriented instructional development (UOID) rejects the idea that technological superiority is a sufficient condition for the adoption of an instructional product. In UOID, the opinions, needs, and perceptions of the potential adopters are seen as the primary forces that influence adoption (Burkman, 1987).

Out of the various aspects concerning the relative advantage of an innovation compared to the old product or practice, studies on the DOI in agricultural field suggest price, quality, risk reduction, and consumer preferences as the most important aspects.

As farmers typically stress over financial issues, they were observed to factor the price of an innovation into their decision to adopt it or not. They were reserved toward new products and practices more expensive than established ones (Senyolo *et al.*, 2018). Yet assuming that higher prices automatically hinder adoption seems too simplistic. Farmers were willing to pay a higher price for better quality, or when they expected the additional cost would lead to an increase in profit (Byrne *et al.*, 2010).

In terms of quality, the innovation would have to be perceived as at least as good as the product it was designed to replace. For instance, one of the reasons why farmers adopted certain nanotechnologies was that nano-sensors performed quality control tasks more efficiently than previous practices (Dasgupta *et al.*, 2015). Farmers were also found to reject products that were qualitatively better but at odds with their values and norms. For example, organic farmers refrain from using antibiotics not due to concerns about their quality but rather because their usage is not compatible with their views.

Due to the already high degrees of risk associated with their field of activity, farmers tend to be otherwise risk-averse. Hence, an innovation's ability to reduce risk was also described as another important criterion influencing adoption intention (Kuehne *et al.*, 2017). As innovations are inherently uncertain and pose certain risk, they may be rejected as long as the old product or practice works. On the other hand, when innovations were perceived as able to reduce existing risk, farmers showed an openness toward them. For example, farmers who are aware their labourers are at risk of injury when performing certain tasks manually may be open to adopting a new machine able to reduce this risk.

Uncertainty about consumer preferences is also a factor that inhibits innovation. When farmers expect consumers will appreciate crops raised organically, they naturally expect that the consumer will be willing to pay a higher price or begin to reject crops produced using more chemicals. Similarly, farmers also assumed that consumers' rejection of meat from animals treated with a large amount of antimicrobials would force them to reduce antimicrobial usage (Visschers *et al.*, 2015).

Complexity

Complexity is the degree to which an innovation is perceived as relatively difficult to understand and use. As Rogers stated, opposite to the other attributes, complexity is negatively correlated with the rate of adoption.

As bureaucracy is among the things farmers complain about most (on account of complexity), an innovation able to reduce this may stand good chances of adoption (Visschers *et al.*, 2015).

The perceptions about the ease of use of an innovation can influence its perceived value and, thereafter, the intention to adopt it. Examples include the ability to use something by oneself, without the assistance of others or expert supervision. Direct effects of ease of use on adoption intention were only seldom reported, but they were described for selected product categories and contexts.

Compatibility

It is the degree to which an innovation is perceived as consistent with the existing values, past experiences, and needs of potential adopters.

The studies suggested that involvement in the issue had a positive influence on DOI. For example, Bourn and Prescott (2002) found that concern for public health influenced the adoption intention of innovations able to protect the environment, particularly in animal husbandry. Problem awareness may be another factor influencing the intention to adopt an innovation. An innovation that helps rectify something seen by the end users as a problem may be more welcome than the absence of such an internalization. It is uncertain to what extent farmers in animal husbandry perceive AMR (antimicrobial resistance) to be a problem, as many studies characterized them as relatively unaware of the risk of AMR and well-aware, respectively.

Trialability

The notion of trialability refers to the ability to try an idea, process or system before making the eventual decision to implement or not to implement the idea, process or system. It is the degree to which an innovation may be experimented with on a limited basis. Because new innovations require investing time, energy and resources, innovations that can be tried before being fully implemented are more readily adopted.

Observability

It is the degree to which the results of an innovation are visible to potential adopters. An innovation more visible will drive communication among the individual's peers and personal networks and will in turn create more positive or negative reactions. Big companies create products and product experiences that are observably different and act as advertisements themselves.

In addition to these characteristics of innovation, Napier (1991) has given one additional attribute "predictability". It refers to the degree of certainty of receiving expected benefits from the adoption of an innovation.

Though the most widely adopted characterization is given by Rogers, in practice, many strange situations arise where in addition to these attributes, we also find other responsible characteristics of innovation. One such situation is when people encounter two or more innovations at a time which are perceived as similar as a whole. Here what attribute does the innovation have which lures the innovator of one innovation. This simply could be the "chance selection" of the innovation. Chance selection of an innovation can be defined as the probability of selecting an innovation among the set of equally perceptible innovations by the people. Thus, one more factor associated with innovations can be identified. *i.e.* chance selection.

Box 4.3

The case of Apple brilliantly utilizing the attributes of innovation in their products is illustrated below:

1. **Relative Advantage** – How improved an innovation is over the previous generation? For example, is the new iPad much better than the previous one? Over a laptop? Over a Microsoft or Amazon product?
2. **Compatibility** – What is the level that an innovation gets assimilated into an individual's life? An iPod is very easy to integrate as Steve Jobs famously said when introducing it to the world, "A thousand songs in your pocket."
3. **Complexity** –Is the innovation too difficult to be used by an individual? Consider touchscreen interfaces vs. multiple menus: We're in a hurry. We're lazy. We don't want to think. Tap a button and experience an app – an iPhone is so intuitive toddlers can quickly use them.
4. **Trialability** – How easily an innovation may be experimented with as it is being adopted? If a user has a hard time using and trying an innovation, the individual will be less likely to adopt it. Is there a sandbox to try it out in first? Think of the Apple retail store experience which is specifically designed to get users interacting with all the devices in a casual way.
5. **Observability** – To what extent an innovation is visible to others? An innovation that is more visible will drive communication among that person's peers and personal networks and will in turn create more positive or negative reactions.

Consider the presence of white headphones that drove the original print campaign for Apple and demonstrated to the world how an individual was listening to an iPod.

Source: http://www.ukessays.com>essays>marketing>the-diffusion-of-innovation

4.10. Social networks in diffusion of innovations

If the innovation diffusion literature since 1970s is reviewed, it could be seen that the emphasis has been on word of mouth, communication, imitation, and what is called the "contagion effect". People infect other people with new ideas or firms infect other firms with new ideas and innovations. That is the major driver of new product growth, of new innovation growth. Social network has a lot of effect on the growth of new products. So that is one of the new fields of study on the rate of growth, and how the social network determines the way of growth and innovations.

The fact is that innovation growth is slow. It is slow because it is being driven by word of mouth, or imitation generally referred to as "social influence". It takes time if one thinks about the movement of the idea or the word of mouth between those who have the innovation and those who don't.

Adopting innovation is also risky. It is not certain whether the innovation will catch on. The risk does not exist if the technology is mature. But if a new technology is adopted, it is not known whether it is going to succeed. It is also not known at that time which technology is going to dominate, which is going to be the industry standard, and so people have to wait. Therefore, it takes time for the product to succeed.

In the absence of formal institutions, social networks are an important mechanism for diffusing information. Social networks are of particular importance for agricultural households in the developing regions of the world. These programmes using social networks implicitly assume that farmers' social networks reinforce extension messages and increase the uptake of technologies. They may be more cost-effective than direct extension visits to farmers in their fields; however, vulnerable or otherwise less-influential community members may be less likely to benefit due to the impact of social norms or the gender composition of the targeted groups.

4.11. Barriers to Adoption of Innovations

Similar to the characteristics encouraging adoptions, it is possible there could also be certain barriers to adoption of an innovation. Ram and Sheth (1989) have put such barriers under three heads- value barrier, usage barrier and risk barrier.

1. Value Barrier

The value barrier refers to a product's lack of relative advantage compared to its substitutes. When cellular phones were launched in India, being costly, it was considered to be the product of exclusive customers. It was also accepted in business markets as a good mobile communication device and hence worth the cost. Since landline phones were already available, most consumers felt that the cost of mobile phone was too high in relation to the value they could get from the landline phones.

Subsequently, the mobile phone market saw phenomenal changes with many manufacturers such as Bharti (Airtel), Hutch, Reliance *etc.*, entering the field and pushing down the price of the handset and offering various schemes that help in reducing the monthly charges. Simultaneously, the mobile phone companies also used various media to convey information on their product and offered various valuable services to their target market, which resulted in mobile phones ultimately replacing landline phones.

2. Usage Barrier

A usage barrier takes place when a product or service is not compatible with the consumer's existing habits. For instance, though we are in the I.T (Information Technology) age, with a reasonable percentage of the consumers being not tech savvy, online shopping has not found wide acceptance.

There is resistance from customers who prefer to visit shops, examine the merchandise, and have their queries attended to by the store personal. Moreover, they feel there is a fun element involved when shopping with friends and relatives. The usage barrier in online shopping occurs because they feel it is not compatible with their desire for visual treat and social interaction involved in shopping when compared to e-shopping.

To a certain extent, marketers are handling this barrier by having opinion leaders (persons who have personal experience with the benefits of e-shopping) communicate about the positive aspects of online shopping. However, with the Covid-19 pandemic, online shopping has become more common.

3. Risk Barrier

A risk barrier refers to the consumer's physical, economic, performance or social risk in adopting an innovation. When cellular phones were introduced, consumers were worried about risks due to reports about physical risks of radiation from frequent and long usage of the mobile phones.

The technological improvements and media communication to educate consumers helped to reduce consumers perception of these risks. Initially, consumers had fears about the price (economic) and performance (complicated software languages) of personal computers (PCs). This risk has been largely reduced by lowering the price of PCs as well as new, easily understandable software packages being made available in the market. Similarly, consumers' social risk of adopting designer wear clothes was overcome through endorsements made of consumers' peer groups who have already used the same.

An effective way of reducing consumer risk is through offering free samples or trial or unique online strategies. Free samples are given during the launch of continuous innovations for products such as tooth paste and detergent powders. Car manufacturers are offering free test drives when new models are launched (Zen Estilo, Corolla, Logan). Auto companies are using a combination of online and off-line strategies to interact with prospective customers. For instance, Toyota has taken the online route to sell its limited edition model of Toyota Corolla. Internet research has become an important part during the purchasing process of automobiles. The company feels that using tools like microsites, direct mailers through the internet, SMS campaigns and having set up a back office operation which will provide the link to its dealers will help the customer in his/her internet research.

Travel portals are now partnering retail chains to increase their volumes by having their presence at such high footfall areas. For instance, make my trip. com had set up a separate counter at Spencer's after receiving an encouraging response for a similar pilot study done at Subhiksha. So apart from flight/ hotel bookings which can be done through internet interface, complete holiday packages can be offered through face to face interaction.

Some marketers use their distribution network by offering off-line transaction services to customers. Yatra online has tied up with Hughes, a broadband satellite network provider to offer its travel related services to customers particularly in smaller cities.

4.12. Rethinking Adopter Categories

The modern way of looking at adopter categories is to divide the market into 'early' and 'main'. The early market in the old categories was either innovators or early adopters; and the main market is usually divided into two segments: early maturity and late maturity. There is also the lag of those who come behind but marketing professionals are not very interested in terms of their marketing efforts. The main question is between the early market and the main market. There are major differences between them.

The early market is more willing to take risks. They are usually wealthier. The risk of adopting a new innovation (is not only financial), and part of it is just the innovation failure. Those are the early adopter category. They will adopt the early market in spite of the inherited risks of early innovation. The main market is not that forgiving. The main market wants the perfect product. While the early market might still be interested in how the product does in the technology itself, the main market is interested only in the functionality. If it functions well, they will buy it. If not, they will not.

There might be a partial break in communication if the early market still talks technology not functionality, such as talking about the RAMs and ROMs, which is not of interest to the 'main market'. If early market start talking about functionality, then it is problematic. It is not the fact that they are not talking to each other but the fact that they are talking different language. But if they stop talking about 'how' instead of 'what' it does for the consumer, then they are still being influenced by word of mouth from the 'early market'.

That would really drive the growth. In the main market, the consumers want someone to tell them that product works. They don't want inherent risks. They want to minimize the risk. What matters is the fact that it works well and the value proposition is there in terms of the price and the performance.

In general, when we look at innovation of successive generations of the same technology, we don't find acceleration, which is surprising. When we look at portable music devices, the original tape, small tape, Walkman, Discman, and then the MP3 player which Sony completely failed in, we don't find acceleration between those successive generations although it was decades between one and the next. It's surprising because we are used to think that there is acceleration in everything we see. Life is faster and the acceleration is still faster.

What we see in acceleration is, first of all, before and after World War II, there is a major difference. We see differences in technologies which require lots of infrastructure. So if we look at electricity, telephones, old fixed line telephones, they required infrastructure and that's why it's been very slow to begin with. When we think about MP3 players, they require infrastructure in the sense of having enough songs and media to go around, but that's basically it. Once technology improves, it diffuses the infrastructure. That's not the major problem, but still, it is seen that they diffuse at about the same rate.

4.13. What is Wrong with Diffusion of Innovation Theory?

Roger's Diffusion of innovations model has been applied in over 5000 studies. It has been well accepted and used in agricultural extension, certainly up to the 1980s when the transfer of technology approach was still prevalent. But even with this success, there have been a number of criticisms.

The four main categories of criticisms of diffusion theory are: pro-innovation bias, individual-blame bias, recall problem, and the issue of equality.

The **pro-innovation bias** implies that it is beneficial for an innovation to be diffused and adopted by all farmers. In reality, farmers only adopt innovations that are appropriate to their context. *e.g.*, Rubber Board did not allow intercropping in rubber gardens earlier if the rubber growers have to avail the extension services of Rubber Board.

The **individual-blame bias**- This occurs when the blame for non-adoption is solely put on the farmers. Sometimes, it could be the fault of the extension practitioner on account of poor or unclear communication of the innovation or just that the innovation is not appropriate

The third criticism is around the **recall problem**. This occurs when adopters of an innovation cannot accurately recall the exact time they adopted the innovation, let alone when they may have first heard about it. This affects the accuracy of the adoption related data and the diffusion curve.

The fourth and major criticism is the **issue of equality**. This relates to the socio-economic benefits of an innovation not being equally distributed through a given population and the consequent widening of the socio-economic gap due to the adoption of the innovation.

Another challenge to the Diffusion of innovations theory comes from Geoffrey Moore who wrote the book Crossing the chasm (2002). He suggests that the adoption curve is not continuous for discontinuous technologies, where adopters are required to substantially change their behaviour or to modify other products and services they use.

Overall, the DOl tradition has sought to explain individual adoption decisions or intentions to adopt. These decisions concern well-defined innovations like a TV set or use of a pesticide among farmers and the adoption population is relatively homogeneous and has well defined boundaries.

The six assumptions of the DOl model are:

1. The technology moves in a discrete package from an independent innovator to the adopter through a constant social "ether" called a diffusion arena
2. The adopter's choice to adopt forms an atomic, isolated decision, which is shaped by the push and pull factors
3. The decision to adopt follows a rational calculus that is based on observed technological characteristics, and other relevant information made available to the adopter through communication channels
4. The diffusion process evolves through distinct stages, which are determined by the push and pull forces and are distinguishable by changes in the adoption rate
5. The diffusion process has neither feedback, nor any "effective" history
6. Time scales are relatively short and the diffusion history is not important.

Lyytinen and Damsgaard (2021) have questioned the usefulness and applicability of DOl to explain the diffusion of a complex, standard-based and networked information technology. The DOl theory integrates two supplementary modes of explanation: the supply-push and the demand-pull theories.

Supply-push theories reckon that specific features of the innovation cause the networked IT diffusion like its functionality, or the standards that enable its use. Networked IT is thus portrayed as a technological fix for organizations' supply- chain problems. Push factors include features of technology, and channels of communication.

The demand-pull theories explain networked IT diffusion by a growing demand for organizational coordination. Organizations need to improve their internal operations, and change their market positions by applying technical knowledge. Pull is determined by adopter's rationale choices.

Rogers' diffusion studies provide two important guidelines for driving the rate of diffusion. They are:

- ☆ Examining social networks (Rogers was not referring to Facebook or LinkedIn, but rather "real life" social networks) and finding highly respected individuals and working with them to create desire for an innovation

- ☆ Determining a representative group of desired users and "injecting" the innovation into that group to gain positive feedback, case studies, *etc.* to help make the decision making process easier for other would-be early adopters.

4.14. Studies Defying Rogers' DOI Theory: Implications

DOI theory dominated the research in Agricultural Extension all over the world for almost half a century, barring certain criticisms posed. Some Extension researchers have indicated that technology adoption in agriculture may not always follow the classical adopter category- wise progression as indicated in the DOI theory proposed by Everett M Rogers.

Though extension services, in general, follow an individual farmer centric approach in promoting new technologies in farming, group approach is found to be more effective in certain contexts such as adoption of IPM. There are also contexts in which group level adoption is needed where social participation and group dynamic skills are found vital, as in the case of FFS.

Some studies defying Rogers' DOI theory is presented below:

Peshin *et al.* (2009) have indicated that DOI theory is not considered adequate to manage the process of dissemination of IPM technology. IPM is not amenable to the limited version of DOI theory. Diffusion of IPM requires educating the farmers for its adoption and it must deal with farmers' needs, perceptions, constraints, objectives and its complexity demands. It requires a clear understanding about the IPM tactics, which vary from crop to crop and area to area. The DOI research has to give up the 'ex-post-facto' design, which has been a prisoner of socio-economic factors influencing the adoption of innovation and in correlating the effects to these factors. The diffusion research should employ 'action research' design to study the IPM implementation and feed the results to develop farmer acceptable IPM system.

Hossain and Quaddus (2010) had studied the adoption diffusion model of Radio Frequency Identification (RFID) based livestock management in Australia and concluded that for pursuing the maximum benefits, mere adoption of RFID is not sufficient. It is necessary to institutionalise this technology into routine operations and practices and extend its use to different applications, where possible, suitable and profitable. The continued and extended use behaviour of the technology is different from and possibly more important than its initial adoption. Many RFID adopters are initially driven by mandatory pressure and later choose different level of use depending on their internal judgment. DOI theory is silent, not explaining how adopters go to the next stage of diffusion and to further innovation applications. DOI theory is primarily based on individual level adoption decisions. But RFID adoption needs to be examined from government and organisational perspectives, because it involves many policies and requires modification of organisational structures and operations.

Using three variations of dynamic logistic model, Arora and Bansal (2011) analysed the impact of certain economic factors like seed prices, technological development, and cotton prices on the diffusion of Bt cotton in India. The study used panel data set covering nine major cotton growing states over the years 2002 to 2008 and found that although seed prices were significant in impacting the diffusion rates, its impact was limited. The study also sheds light on the significance of technological development in impacting diffusion rates.

Although professionals have pointed out artificial insemination (AI) as an emerging technology of socio-economic importance in the livestock sector in India, the ground realities or practices about AI is entirely different. With regard to adoption of AI, Gandhi *et al.* (1998) reported that in spite of huge infrastructure and budgetary provisions over the years, only about 10 per cent of bovine population in India was covered through AI.

Rathod *et al.* (2017) made an attempt to study the adoption status of AI using data collected from 360 dairy farmers in four states of North India. The adoption status of AI pointed out that, majority of the respondents had adopted AI for more than nine years, which was mostly partial in nature. The study concluded that since majority of the dairy farmers were in partial adoption category, a need-based long-run study under field conditions must be undertaken. The study recommends for participatory technology generation and transfer approach with adequate representation of all categories of farmers so that their feedback can be taken into consideration.

Based on a long-term action research study on bio management of Coconut Rhinoceros Beetle in Kerala, Anithakumari (2022) highlights the need for a more nuanced approach to look at technology dissemination and adoption. She concludes that the consequences of adoption can be private, public, or both, as the case may be.

Readiness to adopt technology by forcing an innovation may have to be modernised for jointly evolving and playing critical roles in the use of such innovation. The social impact/public consequences of technology dissemination and adoption call for impact analysis on a system basis. The decisions for adopting technologies follow different pathways in individual and community approaches. The dimensions of access, equity, institutional structures, and common investment by community organizations require further study and analysis.

4.15. Issues Related to Innovation and Diffusion Studies

Some of the prominent researchers and their works are briefly reviewed to highlight the emerging issues related to innovation and diffusion research.

Rogers (1971) identified seven major and six minor diffusion research traditions in terms of the number of empirical studies conducted in each. The traditions he identified were:

Major Traditions	*Minor Traditions*
Anthropology	Agricultural Economics
Early Sociology	Geography
Rural Sociology	General Economics
Education	Speech
Medical Sociology	General Sociology
Communication	Psychology
Marketing	

While identification of these traditions in terms of the number of studies conducted is of historical interest, it sheds little light on their substantive differences and similarities.

Kelly and Kranzberg (1978) concluded that the field of diffusion studies is still so fragmented by specialized concerns that even the dimensions of an adequate general theory are poorly demarcated. They indicated that diffusion research has been conducted under three broad perspectives- a) Geographer's tradition, b) Economic perspective and c) Social-psychological tradition.

Geographers are concerned with spatial diffusion, specifically the relationship between innovativeness of adopters and their relative position in physical space. Hagerstrand (1968) demonstrated that the diffusion of innovations exhibits three states of growth in the spatial distribution of adopters. Brown (1969) distinguishes among macro-, meso-, and micro-scale diffusion within an urban area.

Grilliches (1957) and Mansfield (1961) were concerned with the relationship of "profitability" or "payoff" to diffusion. Mansfield (1961) offered evidence that more profitable innovations are adopted more rapidly than less profitable ones.

Two separate studies heuristically illuminate social, cultural, and perhaps psychological components of resistance to change in the diffusion of technological innovation. Spicer (1953) studied fifteen cross-cultural cases of successful and unsuccessful attempts to introduce new ideas and methods in agriculture, industry, and medicine. He concludes that the change agent, coming from another culture, must be careful to operate through existing channels of communication and cooperation and must understand the relationship of his position as an outsider to the members of a given social system. Bright (1964) studying resistance to innovation as a problem of management, came up with findings similar to Spicer's. According to Bright, resistance is associated with the degree to which institutions and individuals feel themselves threatened by change.

Brown (1981) in his book titled 'Innovation Diffusion: A new perspective' clearly portrays three aspects of the diffusion of innovations: the suppliers' behaviour, the adoption process and the impacts. The book presents detailed explorations of diffusion processes (rather than references to empirical work) in many settings and under a variety of conditions. This information is used to enrich the theoretical development rather than simply to illustrate it.

Valentine and Davis (1999) remarked that interventions designed to use interpersonal communication for promoting behavioural change are often referred to as peer influence, interpersonal counselling, outreach or peer networks. It is assumed that some individuals will act as role models for others, and these role models act as opinion leaders within their communities and can be important determinants of rapid and sustained behavioural change.

Sahal (1981) in his book 'Patterns of Technological innovation' has limited diffusion to technological innovations and used Economics coupled with technological forecasting as a disciplinary base to explain the concept of diffusion. It has important case histories and extensive data on how breakthroughs become the seeds of whole new systems.

Clayton M Christensen (1997) in his bestselling book 'The Innovator's Dilemma' questioned as to why successful companies often can't capitalize on the opportunities brought about by major changes in their market. He has mentioned two types of technologies, *i.e.* sustaining and disrupting technologies. Sustaining technologies are those that improve product performance or that involve improving an already established product. Disruptive technologies, on the other hand, are innovations that result in worse product performance especially in the near term. He explains that disruptive innovations cause problems as they don't initially satisfy the demands of even the high end market; hence the larger companies ignore them until they become more profitable. These technologies later even surpass the sustainable technologies and satisfy market demand at lower prices, thus leaving behind the larger companies that didn't invest in disruptive technologies beforehand.

According to Kotler (2009), diffusion of innovation research and concepts offer a tremendous amount of insight for social marketers to use in designing their programmes, though very little active discussion about this is evident in social marketing circles. Social diffusion offers one of the most robust theories for taking innovations in ideas, behaviour and practice to scale.

There are lessons learned over the years that can be used to reconceptualize social marketing strategy to promote diffusion of products and service use as well as adoption of new behaviours. The weak ties between innovators and early adopters have to be targeted through identifying the 'boundary spanners' who interact with multiple networks. Spanable social distances between groups may be created through various means including using social media such as social network sites and blogs and convening meetings of the 'unlike' rather than the usual host of agents.

Geoffrey A Moore (2014) in his book 'Crossing the Chasm', explained that to ensure an innovation is taking hold, the chasm must be crossed. The gap between the early adopters and the late adopters when we look at the bell shaped diffusion curve is termed as the chasm. If the early adopters are able to cross this chasm and help in bridging this critical point, then the innovation reaches the tipping point, thus allowing the curve to rise up where the majority adopt the innovation and then sink again. He also stated that the population lying before the chasm are the people who want the newest things while the population lying after the chasm are those who want solutions and convenience. The real challenge lies in crossing the chasm, *i.e.*, winning the heart of the 84 per cent who believe in solutions and convenience. Thus, knowing the right time of chasm crossing will certainly help in planning for better adoption of any innovation.

REFERENCES

Ajzen, I (1991) The Theory of Planned Behaviour, Organizational Behaviour and Human Decision Processes, 5 (2): 179-211

Ajzen, I., and Fishbein, M. (1977). Attitude-behaviour relations: A theoretical analysis and review of empirical research. *Psychological Bulletin, 84*(5), 888–918. https://doi.org/10.1037/0033-2909.84.5.888

Anitha Kumari, P (2022) Defying Adopter Categories among Farmers: Lessons for Extension, AESA Blog 168

Arora, Anchal and Bansal, Sangeeta (2011) Diffusion of Bt Cotton in India: Impact of Seed Prices and Technological Development 2011 Annual Meeting, July 24-26, 2011, Pittsburgh, Pennsylvania 104154, Agricultural and Applied Economics Association

Bass, F (1969) A New Product Growth Model for Consumer Durables. Management Science 15: 15-21.

Braak J V (2001) Individual characteristics influencing teachers' class use of computers. Journal of Educational Computing Research 25: 141-157.

Berkhout, G. (2017) Cyclic innovation model (CIM). Available at http://www.aj-berkhout.com/Cyclic per cent 20Innovation per cent 20Model/

Brown, L A (1981) Innovation Diffusion: A new Perspective, Methuen, London

Bourn, D. and Prescott, J. (2002). A comparison of the nutritional value, sensory qualities, and food safety of organically and conventionally produced foods. Critical Reviews in Food Science and Nutrition, 42(1), 1-34

Burkman E (1987) Factors Affecting Utilization. In R. M. Gagne (ed.) Instructional technology: Foundations. pp 91-101, Lawrence Erlbaum, Hillsdale, New Jersey

Byrne, F., Robertson, M. J., Bathgate, A., Hoque, Z. (2010). Factors influencing potential scale of adoption of a perennial pasture in a mixed crop-livestock farming system. Agricultural Systems, 103, 453-462

Barbara Wejnert (2002) Integrating models of diffusion of innovations: A conceptual framework. Annual Review of Sociology 28:297-326. 10.1146/annurev.soc.28.110601.141051.

Cochrane, W. W (1989) The Curse of American Agricultural Abundance: A Sustainable Solution; University of Nebraska Press: Lincoln, NE, 2003.

Chamala S, Van Den Ban A W and Roling Niels (1980) A new look at adopter categories and an alternative proposal for target grouping of farm community. Ind J of Ext Edu 16: 1-18

Dasgupta, N., Ranjan, S., Mundekkad, D., Ramalingam, C., Shanker, R., Kumar, A. (2015) Nanotechnology in agro-food: From field to plate. Food Research International, 69, 381-400.

Davis F D, Bagozzi R P and Warshaw R (1989) User acceptance of computer technology: a comparison of two theoretical models. Management Science 35: 982-1003.

Dearing, J W (2019) Diffusion of Innovations: From Rogers to Today, University of Wisconsin, Madison

Dodgson M and Besssant J (1996) Effective Innovation Policy: A New Approach. pp 15-19. International Thompson Business Press, London

Eastlick M A and Lotz S (1999) Profiling potential adopters and non-adopters of an interactive medium. International Journal of Management, 27 (6): 209-223.

Gandhi, R S, Singh A, Joshi B K (1998) Current scenario and future prospects of sustainable animal production in India, Indian Dairyman 50:35-43

Gartner Research. (2018) Gartner hype cycle. Accessed from https://www.gartner.com/technology/research/methodologies/hype-cycle.jsp

Gatington, H and Robertson, T S (1985) A propositional inventory for new diffusion research, Journal of Consumer Research, 11 (March): 849-867

Gazbar Y (2013) Models of diffusion, adoption, innovation and acceptance of new technology, and social communication. Interdisciplinary Journal of Contemporary Research in Business 4: 811-813

Hossain, M A and Quaddus, M (2010) An Adoption Diffusion Model of RFID-based Livestock Management System in Australia, Human Benefit through the Diffusion of Information Systems Design Science Research, 30 March, 2010

Kelly Patrick and Kranzberg, Melvin (1978) Technological innovation: A Critical Review of Current Knowledge

Klerkx, L., B C Van Mierio and Leeuwis (2012) Evolution of systems approaches to agricultural innovation: Concepts, analysis and interventions, In Farming Systems Research into the 21st Century: The New Dynamic (pp.457-483), Springer, Netherlands

Kotler, P (2009) Social models of marketing diffusion, https://socialmarketing.blogs.con.2009

Kuehne, G., Llewellyn, R., Pannell, D. J., Wilkinson, R., Dolling, P., Ouzman, J., Ewing, M. (2017) Predicting farmer uptake of new agricultural practices. Agricultural Systems, 156, 115-125

Lyytinen Kalle and Damsgaard (2021) Jan What's so wrong with the Diffusion of Innovation Theory? Enablers of Change 7 June, 2021

https://www.enablersofchange.com.au › whats-so-wron.

Maloney (2010). The secret to accelerating diffusion of innovation: The 16 per cent rule explained.

Available at https://innovateordie.com.au/2010/05/10/the-secret-to-accelerating-diffusionof-innovation-the-16-rule-explained/

Moore A Geoffrey (2014) Crossing the chasm. In: Marketing and selling disruptive products to mainstream customers. Third edition. Harper Collins Publishers

Pankaj Kumar and Prabhjot Kaur (2014) An Analysis of Theories of Diffusion, American International Journal of Research in Humanities, Arts and Social Sciences, 6(2), March-May, 2014, pp. 178- 185

Peshin R, Bandral RS, Zhang WJ, Wilson L, Dhawan AK (2009) Integrated pest management: a global overview of history, programs and adoption. Pp 1–49. In: Peshin R, Dhawan AK (Eds) Integrated pest management: innovation-development process, vol.1. Springer, Berlin. doi: 10.1007/978-1-4020-8992-3_1

Ram, S and Sheth, J N (1989) Consumer Resistance to Innovations: The Marketing Problem and its solutions, Journal of Consumer Marketing, Vol 6, No,2 pp 5-14

Rathod, Praksashkumar, Mahesh Chander and Chethan Sharma G (2017) Adoption status of artificial insemination in Indian dairy sector: application of multinomial logit model, Journal of Applied Animal Research, Vol 45, Issue 1

Rogers E M (1983) Diffusion of innovations. pp 1-268. The Free Press, New York

Rogers E M (2003) Diffusion of innovations (5th Edn) Simon and Schuster, 2003

Rogers E M and Shoemaker F F (1971) Communication of innovations: a cross-cultural approach. Free Press, New York

Sahal, Devendra (1981) Patterns of Technological Innovation, Addison Wesley Publishing Co, New York

Segal H P (1994) Future imperfect: The mixed blessing of technology in America, The University of Massachusetts Press, Amherst.

Siddaramaiah B S and Nithyashree D A (2005) A new method of adopter categorization – PRIM(E) Model. Ind J of Ext Edu 41: 1-6

Silk, K. J., Hurley, A., Pace, K., Maloney, E. K., Lapinski, M. (2014) A diffusion of innovations approach to understand stakeholder perceptions of renewable energy initiatives. Science Communication, 36, 646-669.

Senyolo, M. P., Long, T. B., Blok, V., Omta, O. (2018) How the characteristics of innovations impact their adoption: An exploration of climate-smart agricultural innovations in South Africa. Journal of Cleaner Production, 172, 3825-3840.

Sulaiman, R.V. and Hall A.J. (2002) An innovation system perspective on the restructuring of agricultural extension- evidence from India, Outlook on Agriculture 30(4): 235-243.

Taylor S and Todd P A (1995) Understanding information technology usage: a test of competing models. Information Systems Research 6: 144-176. Tessmer M (1990) Environmental analysis: a neglected stage of instructional design. Educational Technology Research and Development 38: 55-64

Visschers, V. H. M., Iten, D. M., Riklin, A., Hartmann, S., Sidler, X., Siegrist, M. (2014) Swiss pig farmers' perception and usage of antibiotics during the fattening period. Livestock Science, 162, 223-232.

Valente, T. W., Rogers, E. M. (1995) The origins and development of the diffusion of innovations paradigm as an example of scientific growth. Science Communication, 16, 242-273.

Valentine, T W and Davis Rebecca. L (1999) Accelerating the Diffusion of Innovations using Opinion leaders, Annals of the American Academy of Political and Social Science, Vol 566, p 55-67

Wilkening, E A (1953) Adoption of Improved Farm Practices as related to Family factors, Wisconsin Experiment Station Research Bulletin. 1283, Wisconsin

Chapter 5

Disruptive Innovations in Agriculture

Disruptive innovation refers to innovation that creates a new market and value network and enters at the bottom of an existing market and eventually displaces established markets, products and services. It is the process by which a smaller company, usually with fewer resources, moves upmarket and challenges larger, established businesses.

It is used to denote innovation that transforms expensive or highly sophisticated products or services, previously accessible to a high-end or more-skilled segment of consumers, to those that are more affordable and accessible to a broader population. This transformation disrupts the market by displacing long-standing, established competitors.

Disruptive innovation refers to the use of technology that upsets a structure, as opposed to "disruptive technology", which refers to the technology itself. It requires enabling technology, an innovative business model, and a coherent value network. Disruptive technology refers to any enhanced or completely new technology that replaces and disrupts an existing technology, rendering it obsolete. It is designed to succeed similar technology that is already in use.

Clayton Christensen, a Professor at the Harvard Business School, popularized the idea of disruptive innovation in his 2003 book 'The Innovator's

Solution', which was a follow up to his earlier book 'The Innovator's Dilemma' published in 1997. He described "disruptive technology " (DT) as one that displaces an established technology and shakes up the industry or a ground-breaking product that creates a completely new industry.

5.1. Types of Innovation

According to Christensen, there are three types of innovation. They are:

1. **Sustaining innovation**, in which a company creates better products to sell for higher profits to its best customers. Science has demonstrated that, through the process of evolution and natural selection, organisms gradually improve and develop in complexity to be more competitive. The same holds true in business, with 'sustaining innovation', where products and services are gradually refined and improved over time.
2. **Low-end disruption**, in which a company enters at the bottom of an existing market and offers a lower-priced product with acceptable performance, ultimately capturing the customers of the competitors.

   An example of a low-end disruption is the rise of retail medical clinics in the healthcare space. Similarly, a low-end disruption emerging in the real estate construction market is 3D printing technology. 3D printing has entered at the bottom of the real estate construction market, claiming the lowest segment: people who need homes that are "good enough."
3. **New-market disruption**, in which a company creates and claims a new segment in an existing market by catering to an underserved customer base, slowly improving in quality until the incumbent businesses' products are obsolete.

The factor that sets new-market disruption apart from low-end disruption is its focus on an audience that does not yet exist in the market. Offering a more cost-effective, simple, or accessible product effectively creates a new market segment.

The iPhone is a classic example of disruptive innovation, where a new product is introduced and changes the entire market.

Low-end and new-market disruption are examples of disruptive innovation, differentiated by their relationships with the existing market.

5.2. Time Line of Major Works on Disruptive Innovations

Disruptive innovations are not breakthrough technologies that make good products better, rather they are innovations that make products and services more accessible and affordable, thereby making them available to a larger population. The time line of disruptive innovations is presented below:

Year	*Disruptive Innovation and Key Features*
1995	'Disruptive technologies: Catching the wave' by Brower and Christensen published in Harvard Business Review reinforce technology as key to disruption.
1997	'The Innovator's Dilemma' by Christensen popularised the theory of disruptive innovation. The book explains why disruptive innovations are difficult for large organisations.
2003	'The Innovator's Solution' by Christensen and Raynor highlights the role of business models in disruption. New market disruption is distinguished from low end disruption.
2004	The modularity theory of Christensen offers valuable insights into the spread of innovation. Modularity has been heralded as an organisational and technical architecture that enhances incremental and modular innovation.
2006	'Disruptive innovation for social change' by Christensen *et al.*, published in Harvard Business Review highlights the role of disruptive innovation in impacting social change. How personal computers emerged as a disruptive product innovation is well explained.
2007	The Christensen Institute is created to further develop the Theory of Disruptive innovation.
2008	'The Innovator's Prescription: A Disruptive solution for Healthcare' by Christensen provides insights into disruption in healthcare.
2011	Christensen *et al.*, published 'The Innovator's DNA: Mastering the Five Skills of Disruptive Innovators', which outlines five discovery skills that distinguish innovative entrepreneurs from ordinary managers: Associating, Questioning, Observing, Networking, and Experimenting.
2013	The Christensen Institute publishes a paper introducing the 'Theory of Hybrids' relating to sustaining innovations that combine existing and disruptive technologies
2016	Hall *et al.* in their book 'Competing against luck: The story of Innovation and customer choice' highlights the need for companies to transform innovation from a game of chance to a planned game.

5.3. Requirements for Disruptive Innovation

Disruptive innovation requires access to ignored or overlooked markets and technology that can transform a product into a more accessible and affordable one. To be disruptive, the network of partners, suppliers, contractors, distributors and other stakeholders must also benefit from the new, disruptive business model. The core requirements include:

- ☆ **Enabling Technology**: Specific to disruptive innovation, enabling technology makes possible the affordability and availability of a product to a broader market. In business, enabling technology is defined as the technologies and innovations that substantially change or improve processes or how people do things. Basically, the speed with which a market can be disrupted depends on how quickly the technology is developed and subsequently improved upon.
- ☆ **Innovative Business Model**: The innovative business model uses innovations to target new or bottom-tier customers. These segments generally do not drive profits for established companies because they either could not afford them or the products were too sophisticated for use by the general public.
- ☆ **Coherent Value Network:** This includes the upstream and downstream business partners that benefit from a successful disruption. The distributors, suppliers, and vendors may require process changes or reorganization to adapt or conform to the new business model. Members of the network must subscribe to the new business model to prevent failure.

Because disruptive technology is new, it has certain advantages, enhancements and functionalities over competitors. For example, cloud computing serves as a disruptive technology for in-house servers and software solutions. It has slowly been adopted by organisations and individuals with the main objective of completely removing traditional computing.

Disruptive technologies are shaping various sectors and also helping them to achieve phenomenal growth. However, agriculture is one sector, where the producers *i.e.* the farmers are at the bottom of the pyramid (BoP) as far as the benefit sharing from agriculture and associated industries are concerned. The scientists in public and private sector are in the process of developing and introducing such technologies, which coupled with digital technologies have potential to change the way agriculture is being practiced in India.

Lack of awareness of disruptive technology in Indian agriculture is a major problem and steps have already been initiated to overcome this. One such initiative is the formation of Consortium of Researchers for Disruptive Technologies in Agriculture (CDTA) during 2019.

Currently, the agribusiness industry is reshaping the agricultural landscape by transitioning the world away from traditional agricultural production and distribution techniques to a system of "more closely coordinated and better planned linkages between agribusiness firms, farmers, retailers and others in the supply chain".

Box 5.1

In a major initiative to apply disruptive technology in agriculture, a consortium of researchers has been formed with the partnership of three leading institutions in the country with the mandate of turning farming from an 'uncertain' to 'assured' activity.

The platform, the Consortium of Researchers for Disruptive Technologies in Agriculture (CDTA), is a joint venture of the academic-cum-research scientists of IIITMK, Thiruvananthapuram, GB Pant University of Agricultural Sciences and Technology, Pant Nagar (Uttarakhand) and Indian Institute Space Technology, Thiruvananthapuram.

The team will advocate and implement application of technologies like Artificial Intelligence (AI), Data Analytics (DA), Internet of Things (IoT), Cloud Computing (CC), Aerospace Observation and Miniaturised Sensors in the agriculture domain. They together present the untapped potential to overcome challenges facing Indian agriculture and to transform it from an Uncertain to Assured one.

Assured agriculture represents a scenario where individual farmers are guided at every stage of crop growth and provided with timely advisories and physical or fiscal assistance, to ensure beneficial returns for investment; without compromising on biological diversity in the immediate vicinity.

Having seized the need for tapping the tremendous potential of disruptive technology in agriculture, it was considered necessary to create a platform for advocating, creating, applying and sharing the new knowledge for the benefit of the Indian farming sector.

Source: https://www.newsexperts.in › researchers-consortium-for-agriculture

Disruption can take two forms:

1. Substituting existing products, processes and business models: AgriTech can be transformational, first, if it involves displacement of incumbent practices that have been existing in a sector or value chain; second, when there is a complete overhaul or radical change in business operation models or the modus operandi of actors; third, when it creates new markets for the poor, such as those created by digital platforms; and fourth, it enables behavioural change through changing cognitive rationalities, which finally leads to changing the underlying norms and culture of the society.
2. Complementing existing products, processes and business models: This occurs when AgriTech supplements existing products, processes and business models. Therefore, it may lead only to alternations or modification in the modus operandi and partial changes to production

practices. In this case, behavioural change may occur, but it does not change the underlying norms and cultures of society. Thus, a disruption scale exists that ranges from high AgriTech disruption potential under substituting to lower potential under complementing

5.4. Disruptive Technologies in Agriculture

As mentioned, a "disruptive technology" is one that displaces an established technology and shakes up the industry or a ground-breaking product that creates a completely new industry. Agriculture has historically depended on disruptive discoveries and innovations to make big strides. Hybrid seed, mechanisation, dwarf varieties of crops, synthetic pesticides, biotechnology, precision farming, *etc.* are some of the examples of disruptive discoveries and innovations that have transformed farming.

There are four innovations that brought about a change in agriculture in the 20th century. They are- a) Improved Seeds- Use of high yielding varieties and hybrid seeds; b) Mechanisation- Use of Tractors and other farm machineries; c) Fertilisers- Use of urea and other fertilizers; and d) Pesticides- Chemicals derived from hydrocarbons enabled farmers to resist pest attack.

These four innovations allowed more food, feed and fibre to be produced from less land. These innovations disrupted the status quo and created immense benefit to farmers and also consumers.

'Demand' and 'supply' are two deciding factors for making any change in agriculture. Demand for agriculture mainly relates to population growth and changing dietary pattern. The world's population is expected to grow to almost 10 billion by 2050, boosting agricultural demand. Income growth in low and middle income countries would hasten a dietary transition towards higher consumption of meat, fruits and vegetables, relative to that of cereals, requiring commensurate shifts in output.

For supply, we may have to depend upon a number of factors, such as: Labour (Increased out migration), Capital (Increased uncertainties), Natural Resources (Water Scarcity, Climate Change, Degraded Land and Environment) and Technology

The new green revolution is not just about production of food, feed and fibre from agriculture "greening" but it includes taking into account all the activities and outcomes of "food-feed-fibre systems" (F3). This system includes all the elements (environment, people, inputs, processes, infrastructure, institutions, *etc.*) and activities that relate to the production, processing, distribution, preparation and consumption of food, and the output of these activities, including socio-economic and environmental outcomes.

Types of Disruptive Technologies in Agriculture

Farming has been transformed by the new technological innovations in agriculture over the years, and technical advancements have impacted the agriculture industry in a variety of ways. Agriculture is an excellent example of an industry that has undergone a significant technological transformation in recent years. Even then, agriculture is considered the least digitised industry. The use of disruptive innovations in farming has to be promoted on large scale.

The five groups of disruptive technologies in agriculture are presented below:

Disruptive Innovation 1: Digital Agtech

Digital Technology is referred to as the fourth Industrial Revolution. The maximum use of digital technology is being made by Information Communication and Technology (ICT) industry, followed by media and finance, wholesale trade, chemicals, mining, education, entertainment, health care, hospitality and the least by agriculture.

Agriculture is now strongly influenced by the so-called fourth Industrial Revolution digital technologies which started having an impact in the 2010s. The term "agtech," or agricultural technology, has come to represent the milieu of exciting new technologies like drones, sensors and intelligent robots. The new agtech is a set of powerful disruptive technologies that have already started to make a difference to small farmers in Asia. These are joined by tools developed through "fintech" or financial technology to help farmers access credit and markets.

1. "Industry 4.0" is a blend of Advanced analytics, Big Data, Robotics and Automation, Artificial Intelligence, Internet of Things (IoT) and Process Digitisation, across the business value chain.
2. Digital agriculture: "Internet of things" (IoT): Data-Enabled Agriculture in Mobile computing, Data Sensor Input, Satellite and Imagery, Drones and Genetics have resulted in transforming agriculture towards more precision and efficiency.
3. Management Information Systems (MIS): Now a days, in many countries, farmers are planning and forecasting and then making their comparisons relating to food production system between forecast with intervention measures taken now and forecast without intervention measures.
4. Automated Disease Detection Technology: Farmers conversant with this system protect their crop by: a) Taking photos of damaged plant part, b) Matching it with set symptoms of diseases/pest attack and c) Applications as per recommended treatment.
5. Knowledge and Skills: Logistics, Management and Finance, Experiential Learning and Storage are some of the fields where knowledge and skills play an important role in farming.

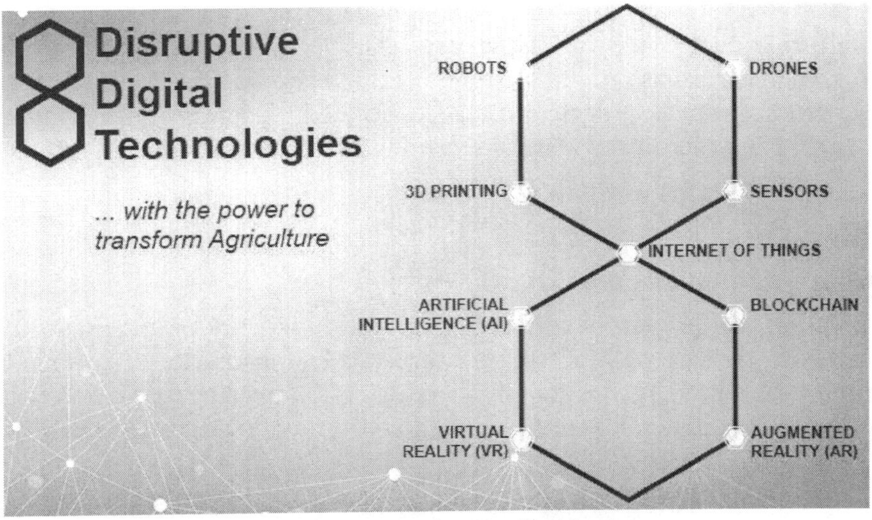

Figure 6: Disruptive Technologies.

6. High Technologies for Precision and Smart Agriculture: High Technologies like Remote Sensing, GIS/GPS, Drone, Fertigation, Biotech, Big Data, Data Analytics, Green House, Robotics *etc.* are being used in different parts of the world, for modern agriculture.
7. Fin-tech Computer programmes and other technology are used to support and enable banking and financial services.

Disruptive Innovation 2: Biotechnology and Nanotechnology

1. Biotechnology- New breeding technologies are developed from the new biology -Beyond GMOs
2. Gene-Editing biotechnologies- Genome edited crops with notable improvements include, soybean, maize, wheat, rice, potato, tomato, peanuts and cotton.
3. Modern biotechnology addresses the challenges and needs of agriculture and society thus resulting in:
   - ☆ Surplus production: more production with less resources, price stability, climate adaptation, new higher-yield crop varieties, *etc.*,
   - ☆ Labour saving production technology
   - ☆ Nutrition and safety
   - ☆ Trade
   - ☆ Negative externalities, *e.g.* pesticide pollution
   - ☆ Environmental stress (drought, floods)
   - ☆ Biotic stress (pests, diseases)

A second disruptive technology is through biotechnology in the form of crop varieties developed using biotechnology, including genetic modification.

Disruptive Innovation 3: Controlled Environment Food Farming Fusing ICT and Biology

1. Plant Factory technology (Urban) – new farming systems. "PFAL" (Plant Factory Artificial Light)
2. Proximity to sources of production in urban areas: freshness, nutrition quality, energy miles - controlled environment farm.

A third disruptive technology has risen in response to urbanisation, climate change and increased demand by consumers to have vegetables grown close by- "Plant Factories with Artificial Light (or PFALs)"- which are essentially enclosed, environment-controlled greenhouses in which vegetables are grown in tiered trays. These plants grown indoors generally have no need to use insecticides and can produce many times more per unit area using LED lights.

Disruptive Innovation 4: Food/Feed/Fibre Technologies (Preventing Product Fraud)

Blockchain-based technology helps increase traceability and transparency in agri-food supply chains. It allows food supply chain partners to connect and share information of interest. This enables food producers to cooperate seamlessly and improves coordination.

A blockchain is a digital ledger of transactions duplicated and distributed across the entire network of computer systems on the blockchain. It consists of associated chains that store data in units called blocks. Blockchain makes it difficult or impossible to change, hack, or cheat the system. It allows farmers to store all their data in one place for easy access, simplifying the entire process and ensuring reliability.

Disruptive Innovation 5: Supply Chain Connectivity-Digital Platforms

Agriculture Information Portal of Ministry of Agriculture and Farmers Welfare, Government of India (http://agrionline.nic.in/) features various platforms such as Farmers' Portal, mKisan Portal, State Specific Agrisnet Portals, Direct Benefit Transfer- DBT, Agriclinics and Agribusiness Centres, Forecast Weather and Agromet Services Information, RKVY Management Information System, SeedNet Portal, Soil Health Card Information System, Industry Specific Projects like Plant Quarantine Information System (PQIS), Fisheries Input management System, National Agriculture Market (NAM), Agriculture Census, DBT in Agriculture Machinery, Farmer/Public Centric Projects, AGMARKNET Portal, Computerised Registration of Pesticides (CROP) and Seed Export Import System *etc.*

Most of these initiatives are aimed at creating platforms without much emphasis on comprehensive, up to date and real time data/information/

knowledge sharing. They appear to have scale in terms of frameworks, but do not have substantial quantities of data that would make a difference.

Even though one can find a number of digital initiatives in agriculture sector, such initiatives are not comparable with the digital disruption models experienced in other sectors. While these are all push based platforms (with few exceptions), the ability to provide customised and personalised information and services all along the value chain is questionable. For a crop based enterprise, farmers may require at least 14 different information and service assistance at multiple stages that would make a difference in informed decision making

Two technologies that have potential to become disruptive technologies are alternative proteins (such as plant-based protein, cellular meat and insect protein), and blockchain technology that safeguards the integrity of food supply from farm to consumer. Blockchains are also spawning the development of techniques to quickly detect food fraud or guarantee food identity.

Digitalization of agriculture will change every part of the agri-food chain. Management of resources throughout the system can become highly optimized, individualized, intelligent and anticipatory. It will function in real time in a hyper-connected way, driven by data. Value chains will become traceable and coordinated at the most detailed level while different fields, crops and animals can be accurately managed to their own optimal prescriptions. Digital agriculture will create systems that are highly productive, anticipatory and adaptable to changes such as those caused by climate change. This, in turn, could lead to greater food security, profitability and sustainability.

In the context of the Sustainable Development Goals (SDGs), digital agriculture has the potential to deliver economic benefits through increased agricultural productivity, cost efficiency and market opportunities, social and cultural benefits through increased communication and inclusivity and environmental benefits through optimized resource use as well as adaptation to climate change.

The potential benefits of digitalizing the agri-food sector are convincing but it will require major transformations of farming systems, rural economies, communities and natural resource management. This will be a challenge and requires a systematic and holistic approach to achieve the full potential benefits.

Digital entrepreneurship involves the transformation of existing businesses through novel digital technologies and the creation of new innovative enterprises characterized by: the use of digital technologies to improve business operations, the invention of new (digital) business models and engaging with customers and stakeholders through new (digital) channels (European Commission, 2013). Globally, there are an increasing number of initiatives to foster digital entrepreneurial activity related to the creation, development and scaling-up of 'digital start-ups', including in the agriculture and food sector.

Box 5.2

The Strategic Policy Forum on Digital Entrepreneurship was set up by the European Commission in 2014 as a powerful think tank composed of business leaders, academia, international organisations and policymakers. During its two year mandate, its objective was to shape a vision for the digital transformation of European industry.

The Strategic Policy Forum reflected on the short and long-term strategy for digital transformation in Europe, advised the Commission on policy issues and actions to encourage digital transformation, and promoted the development of policy by EU countries at national and regional level

In March 2016, the Forum delivered its final policy recommendations, urging action on what it considers the most critical needs of European businesses. This includes the need for cyber-security, high-performing infrastructures, a Single Market for European data, actions to reskill our workforce, and building up innovative ecosystems to take advantage of the digital revolution.

Source: http://single-market-economy.ec.europa.eu>strategy

5.5. Transformative Disruptive Technology Innovations in Agriculture

Emerging technologies that are disruptive in nature have transformational impact on society, economy and environment. The focus is on innovative strategizing, development, standard setting and governance of such technologies keeping in background the lessons drawn from other technological interventions. Out of several new technologies, the technologies presented below are considered powerful transformative technologies.

1. GIS (Geographic Information System) and GPS (Global Positioning System)

Applying GIS software has an important role in precision farming, since fields are location-based. The use of GPS and GIS together allows for real time data collection that acquires an accurate position. This means that farmers can use devices to map where exactly the resources of that farm will be used to increase the resource use as well as increase its efficiency. GPS has also given rise to field mapping which is a system for computer aided field data collection that takes several factors into account such as inventory of crop types, elevation levels, field boundaries, nearby roads, irrigation systems, *etc.*

2. Big Data

Farmers have been managing their land with extension advisories coming from multiple sources. But neither the extension advisories nor the decision-making is based on the microscopic analysis of data from each farm. If such huge data is collected, collated and processed with big data analytics and real time advisories are pushed, then we can realize much talked about personalized advisories. From pre-production (credits, input supplies) to production (varietal to management) and to post-production (processing and marketing), extension systems can harness big data platforms for better and informed decision-making.

The big data analytics in extension will bring significant changes in the personalized, field specific solutions along with pre-production to post-production service needs of farming community. It will lead to higher yields, lower input use per hectare and lower cost of cultivation.

3. Imagery from Satellites

Using satellite images and RADAR to monitor soil properties, crop conditions, and tillage activities enables stakeholders to determine land usage, harvest prediction, observing changes in season, and helping in proposing and implementing a financially and ecologically viable sustainability development policy. Satellite monitoring of the crops allows tracking of the positive and negative changes in the crop development using high resolution satellite images. With this technology one can easily predict the crop yields and also conduct real-time field monitoring.

4. Internet of Things (IoT)

IoT in agriculture uses robots, drones, remote sensors and computer imaging combined with continuously progressing machine learning and analytical tools for monitoring crops, surveying and mapping the fields and provide data to farmers for rational farm management plans to save both time and money.

It enables physical objects to be sensed and controlled remotely, creating opportunities for more direct integration between the physical world and computer-based systems. Data from devices can guide farmers' decisions, helping them farm smarter and safer and adapt more quickly to changing conditions.

Dairy and livestock farmers have been using radio frequency identification (RFID) to enable tracking of individual animals' health and levels of production for quite a number of years. There are many other potential areas for development, including sensor networks to monitor soil and crop conditions, equipment monitoring and automation (self-driven planters/harvesters *etc.*). These devices collect useful data with the help of various existing technologies and then autonomously flow the data between other devices.

5. Cloud Computing

It is an IT paradigm through which users can access shared pools of configurable system resources over the internet. Such a sharing of resources enables coherence and economies of scale. It has become more real with increasing bandwidth, storage, and network speeds. Today, companies and consumers alike are comfortable with storing important information, pictures, and other digital content in the cloud. Start-ups such as Box.com and Dropbox are examples of companies that enable people to store digital assets in the cloud. Software as a Service (SaaS) provide users with Internet-based platforms of services that do not require customers to install and maintain the software on their internal servers.

Some of the specific applications in agriculture are:

- ✰ It can capture information related to all crops grown in the recent past, and thus can help farmers make decisions on what to grow next.
- ✰ The growth of various crops can be monitored in different regions; and at regular intervals.
- ✰ Solutions can be made available for commonplace problems that farmers frequently face.
- ✰ Through the agricultural management information system of cloud computing, farmers can sell their produce directly to the end users/ retailers.

Cloud computing has been a hugely disruptive technology in the business world, displacing many resources that would conventionally have been located in-house or provided as a traditionally hosted service. The next and most radical generation of mobile communications – fifth generation (5G) may radically change the technologies and business models of the mobile telecommunications industry. It will have positive consequence in primary sector like agriculture and extension systems will have to gear up to develop frameworks for best use.

6. Farming Software and Data Over the Internet

Farm management software includes websites, mobile apps, and computer programmes built for the farming sector and helps the farmer make more informed production and operational decisions. Being able to make decisions supported by data, having more accurate farm records, and the possibility of tracking and monitoring farm activities in real-time not only increase profits but also lead to more sustainable farming practices.

7. Drones

They encourage farmers to solve other assorted barriers and receive plenty of benefits through precision agriculture. With the market for agricultural drones catching up in India, UAVs (unmanned aerial vehicles) fill the gap of human error and inefficiency by traditional farming methods. The purpose of adopting drone technology is to exclude any guesswork or ambiguity and instead focus on accurate and reliable information.

The central government has acknowledged the importance of UAVs, machine learning and artificial intelligence with their 'Digital Sky Platform' online. Drone strartups in India have used this opportunity to accomplish better technological capacities.

Drones have various applications in agriculture, but one issue that has received a lot of attention recently is the disappearance of bees. Some species are even on the verge of extinction. This would be devastating, because bees play an important economic function as pollinators, allowing agricultural production to continue at current levels. Fortunately, drones are currently being tested in order to enhance the pollination work that the remaining bees are doing.

8. Robotics and AI

The main goal of agricultural robots is the automation of field activities, saving farmers time on dull, regular tasks so that they could focus on more important meaningful work. Agriculture is slowly moving to automated farming based on the widespread use of mobile and stationary robots. With the power of AI, farmers can track weather changes, precipitation, pests infestations, and more to ensure precise water, fertilizers, and pesticides application based on accurate data.

9. Alternative Energy

There is significant potential for agricultural involvement in the production and consumption of solar, wind, geothermal, and biomass energy. Renewable technologies are now supplying or supplementing many on-farm energy requirements, from water pumping to space heating. This is contributing to greater energy security in agriculture through increased diversity of energy sources, more self-supply of energy, and reduced environmental impact.

10. Vertical Farming and Urban Agriculture

Vertical farming is an energy intensive system of crop production involving integration of multiple technologies such as big data analytics, robotics, internet of things, artificial intelligence *etc.* so that crops can grow well without any agronomic constraint. Traditional farmers have to learn a lot from vertical farmers about how to design and build structures that maximize yield while minimizing waste.

11. Genetic Modification

CRISPR/Cas9 is now being used by scientists to perform genomic "surgeries." By altering or eliminating certain genetic sequences on a strand of DNA, they can identify and alter an organism's genome. CRISPR is now being used to change the intestinal microorganisms of cows in order to manage their size and reduce the amount of methane they emit. Changing the microbiomes of cows in this way could result in more meat being produced with less food. As a result, the industry becomes more efficient and profitable.

5.6. Smart Innovation in Agriculture

Smart innovation in agriculture is referred to as smart farming. It is an emerging concept that refers to managing farms using technologies like IoT, robotics, drones and AI to increase the quantity and quality of products while optimizing the human labour required for production.

Smart Farming solutions leverage digital technology innovations that have the potential to drive positive impact for small scale producer (SSP) entrepreneurs delivered through bundled farmer services and enabled by scalable products.

Some of the features of smart farming are:

- ☆ **Leverage digital technology innovations** with the potential to scale, for example, utilizing a proven technology that has been applied in agriculture in another geography; transferring a technology from another sector such as health, infrastructure or finance; sourcing an entirely new technology;
- ☆ Show potential to improve SSP **incomes, productivity and ability to adapt to climate change;**
- ☆ Address **barriers to scaled adoption of digital services**, such as access, affordability and language and digital literacy. This includes addressing **women's access** to digital services and closing the digital divide between those who have access and marginalized populations who do not;
- ☆ Have the potential to be **bundled** with multiple farmer-facing services into an integrated solution;
- ☆ Are enabled by robust **digital and data technology platforms and services;**
- ☆ **Provide fit-for-purpose solutions** (*e.g.* developed using human-centred design) for SSP enterprises who seek to sell a portion of production to markets;
- ☆ Show **potential to be scaled using a sustainable business model and are affordable and able to provide positive return on investment** at a smallholder farm level through successful pilots, scaling partners, or higher volume production driving down prices.

Some of the recent smart farming solutions are:

AgriBot – Autonomous Tractor

Tractors are now considered an essential agricultural machinery used right from preparing the soil to harvesting crops. It can also be used for transport of farm inputs and produces. Automation in driving technologies has reached agriculture as well with autonomous tractors, freeing up time for farmers to focus on other aspects of agriculture. Polish startup Agribot has developed its eponymous autonomous tractor, the *Agribot,* offering several benefits for the farmer. Apart from saving time, their tractors are fitted with a range of sensors to significantly reduce human contact with chemicals and pesticides during farm operations. It allows for increased capacity, working day and night along with the ability to carry out two agronomic tasks simultaneously.

One Water - Smart Irrigation

The changing seasons have a great impact on agricultural activities. Prolonged drought conditions during the summer months and devastating floods during the rainy season affect crop yields all over the world. Some regions in the world are facing years-long drought conditions with minimal rainfall. As a result, innovations in irrigation are receiving a lot of attention. Smart irrigation provides optimal water delivery to crops while ensuring there is minimal to no wastage in water used for agriculture. Indian startup One Water, guided by their parent company Nascent Info Technologies, develops Internet of things (IoT) based smart irrigation systems for agriculture. Excess water can also cause the destruction of crops and spur the growth of weeds. One Water can sense soil moisture, humidity, and temperature to automatically execute drip irrigation on the farm, saving valuable resources.

Saga Robotics - Autonomous Harvesting

For farms with low mechanization, harvesting is the most labour-intensive activity. Harvesting is not limited to just fields but includes orchards, vineyards, and plantations too. Automation in harvest technologies for agriculture ensures lower stress on labour and higher efficiency in harvesting crops. It also protects the labour from potential accidents. Norwegian startup Saga Robotics develops a modular robotic platform called *Thorvald,* an autonomous farm robot capable of performing several tasks, including harvesting of fruits and vegetables. Farmers can choose from four different types of robots based on their specific needs. All their robots use the ROS software, an open-source meta-operating system for robots.

Desamis - Livestock Monitoring

Cattle, sheep, pigs, goats, and chicken contribute significantly to food security. Cattle are being provided with their own sensors that track their movement and activity, monitor heat in their stomach, check their breath to detect diseases, and more. Japanese startup Desamis developed *U-monitor,* an Artificial Intelligence based smart dairy platform to monitor, detect, and analyze cattle behaviour so that the farmer can make informed decisions. It records and monitors when cattle are feeding, drinking, moving, standing, lying and ruminating. It looks for dysstasia-related diseases using sensors in the stomach of cattle. Moreover, it provides the data in real-time, allowing for immediate analysis and decision-making.

Plastomics - Genetic Editing

It is estimated that by 2050, there will be around 9 billion humans in the world. Presently, almost all arable land in the world is being used for agriculture. This drives innovation in agriculture to significantly increase farm output or farm productivity. One of the most powerful, but controversial, innovations is the genetic manipulation of crop seeds. Techniques like C4 photosynthesis, CRISPR/Cas9 or genomic selection are powerful because of their ability to make crops highly resistant in tough growing conditions. They are controversial because there is little control over the result of editing a gene of a crop, besides the high costs involved for such processes to eventually feed billions. The US-based startup Plastomics is developing a new way to improve crops by making modifications to the plant's genetic traits using its chloroplast rather than its nucleus. It is believed that chloroplast engineering is better since there are many chloroplasts in a cell and hence the traits can be precisely introduced. The biological characteristics of the chloroplast translate into crop products that are better for the planet and more profitable for seed companies and farmers.

5.7. Recent Disruptive Innovations yet to Become Popular

Laser Scarecrows

They are helping farmers to protect their sweetcorn crops from marauding birds. Developed by University of Rhode Island (URI), Professor Rebecca Brown, the new system uses constantly moving beams of green laser light to scare off birds like starlings and red-winged blackbirds before they can tuck into a ripe ear. The automated laser darts over fields up to 600 feet, effectively startling birds and preventing crop destruction.

Bee Vectoring Technology (BVT)

It helps farmers achieve greater yields and fight pests with bees. Ashish Malik has created an agricultural solution for effective pest control. BVT's technology starts in a 'smart hive', where bees dip their bodies in a non-toxic biological fungicide before going off to do their bee business. Bees pass the fungicide along to the plant 95 per cent more efficiently than farmers' sprays, which means only less fungicide is needed in the long run. The proprietary fungicide technology, is known as a "biologic", or a living organism that can help fight off or deter agricultural pests. "Biologics" have become a recent buzzword in agriculture, even though the technology has actually been around for more than 20 years.

Radio Frequency Identification (RFID)

Basically, RFID systems consist of three parts: an RFID tag, an RFID reader, and an antenna. This technology uses digital data, which is encoded in RFID tags and read by a reader via radio waves. RFID is comparable to barcoding. RFID has the advantage over barcodes in that the RFID tag data can be read without being seen, whereas barcodes must be seen so that an optical scanner can read it.

In a farm context, RFID are more efficient and comprehensive because they can be read invisibly and from many feet away in the mud. RFID, unlike barcodes, can be reprogrammed and can store up to 2KB of data, including information like location, date, and time.

RTK (Real-Time Kinematic)

It is a new technique for improving the accuracy of existing GPS signals. It is a technique that uses carrier-based ranging and provides positions that are orders of magnitude more precise than those available through code-based positioning. Differential GPS usually gives about one meter accuracy. RTK is a special form of Differential GPS that gives about one-hundred times greater accuracy. The approach eliminates the vast majority of common GPS mistakes, allowing for centimetre-level precision. RTK improves GPS or satellite tracking in the field using stationary sensors to better estimate farming yields and costs.

With the emergence of agricultural technologies integrated with a robust information and communication technology (ICT) framework that is still evolving in India, there is tremendous potential to positively impact agricultural performance and enhance farmer's income that is still unmet.

- ☆ International Crop Research Institute for Semi-Arid Tropics (ICRISAT), a UN agency, analyse volumes of data on weather forecasts, local rainfall and soil conditions. This data collected gets analysed to develop a 'sowing date' application that tells farmers the right sowing date to maximise their yield. Access to this platform is further simplified by providing information to farmers via SMSs in regional languages.
- ☆ The "uberization of the tractor sector" through a mobility app called "Trringo" enables farmers to rent tractors. This is a unique example of leveraging technology to help farmers use machinery without having to make the large investment of buying tractors.
- ☆ Some leading startups have been providing dairy farm optimization and monitoring services with a special focus on small- and medium-herd farms with the help of tools that leverage the Internet of things, big data, the cloud, mobility, and data analytics to improve milk production, milk procurement, and the cold chain, and to boost animal insurance and farmer payments.
- ☆ Availability of real-time data is one of the most significant opportunities for technological innovation. Also, drones are fast becoming a real green-tech tool. Global research also shows that AI farming will be the main enabling factor in increasing the world's agricultural production capacity to meet the demands of the growing population.

☆ Mission mode schemes/programs launched to propel the development and deployment of frontier disruptive technologies such as blockchain, AI, 3 Dimensional (3D) printing, Quantum Internet of Things (IOT), *etc.* impact society and the economy across the sectors.

5.8. NITI Aayog and Disruptive Innovation

The disruptive models that are changing the landscape of the farmers' ecosystem have been well recognised by the Government and have also necessitated amendments in the farming policies from time-to-time. In current times, startups have penetrated throughout the farming value chain from sourcing quality seeds to providing mechanized tools for farming, logistics support for movement of crops and improving access to new markets. In the recent years, with application of new age technologies like blockchain and AI in the farming ecosystem, entrepreneurs have been able to provide farmers with advanced and futuristic solutions to improve their output. Use of such advanced technologies have also provided the startups with new funding opportunities as they are now trying to concentrate and solve the major challenges that the farmers community will face due to climatic changes.

Blockchain Technology in Agriculture

Blockchain in agriculture can be applied in optimizing resources in crop production, controlling weather crisis, traceability in food supply chain, managing agricultural finance, and simplifying transaction processes. Blockchain's most basic promise for the agriculture industry is that it removes the need for third parties otherwise required to ensure trust within buyer-seller relationships. A transparent and trusted system can be built by putting all the information about agricultural events on a blockchain. Farmers can also get instant data related to the seed quality, climate environment related data, payments, soil moisture, demand and sale price, *etc.* all on a single platform.

FOODSIGN, the flagship solution from Bangalore based TraceX Technologies, is a digital agricultural platform that leverages blockchain. It connects multiple supply chain participants such as farmers, aggregators, food processors and retailers to securely ex- change verifiable and auditable data leading to mutual trust, accountability and transparency among the participating actors, eventually reducing food waste and increasing consumer trust. The stakeholders can trace the pre- and post-harvest journey of the produce using a combination of multilingual An- droid app, web application, application programming interface (API) and IoT devices.

ICRISAT, in partnership with Eleven 01, India's native blockchain platform and KHETHINEXT, an incubatee of ICRISAT's innovation hub has deployed blockchain technology in agriculture to increase productivity. KHETHINEXT has developed a decentralized application that provide farmers with mobile-based expert advisory services, ground support for logistics and blockchain-based clarity into financial transactions and supply chain monitoring. This can empower smallholder farmers with information regarding farm inputs, produce demand/supply, market rates, *etc.*, eliminating intermediaries and enabling the farmers with better decision-making powers.

Blockchain for Natural Farming: Lack of quality produce and traceability have hampered India's food exports, apart from disincentivising growers. Blockchain technology can help bridge this gap. To pilot such an initiative, the Agriculture Vertical, in consultation with the Frontier Technologies Vertical, has been working with Neural Blocks, a start-up that provides blockchain-based solutions. A pilot project has been launched in Himachal Pradesh, with the support of the State Government, in providing data. Apple has been chosen as the first crop.

Artificial Intelligence (AI) in Agriculture

Automation in agriculture is an emerging activity across the globe. In recent times, AI has been witnessing a lot of direct application in farming. AI powered solutions not only enable farmers to do more with less and reduce wastage, but also improve quality and ensure faster go-to-market for crops.

Cognitive computing has become the most disruptive technology in agricultural services. Microsoft is working with selected farmers in Andhra Pradesh to promote agriculture, land and fertiliser advisory services. Some of the uses of AI innovations are:

1. AI technology innovations in agriculture aid in the detection of plant disease, pests, and poor agricultural nutrition. AI sensors can detect and target weeds, then determine the best herbicide to use in the area.
2. AI-based Robots for Harvesting- These machines use sensor fusion, machine vision, and artificial intelligence models to identify the location of the harvestable produce and help pick the right fruits.
3. AI for Predictive Analytics- ICRISAT scientists employed a predictive analytics technique to determine the best time to sow the seeds for optimal yield. In addition to a 7-day weather forecast, it provides information on soil health and fertilizer recommendations.

Agriculture is one of the focus areas of the National AI strategy document released by NITI Aayog in 2018, which asserts that AI and related technologies have the potential to improve productivity and efficiency across all stages of the agricultural value chain and can help increase farmer income and farm productivity by reducing wastage.

The State governments have also partnered with major technology companies to roll out AI-powered solutions for agriculture. The Government of Karnataka signed a Memorandum of Understanding (MoU) with Microsoft to set up a system for improved price forecasting and to provide farmers with advice

Box 5.3

Precision agriculture startup Fasal uses AI, IoT, and analytics to remove the guesswork from farming, enabling thousands of horticulture farmers to manage their crops remotely and profitably.

In early 2017, Ananda Verma along with his friend Shailendra Tiwari bought a small farmland near Mysore to grow coloured capsicum (bell pepper) as a weekend getaway from their big-city life. Though both have a farming background, they realised the problems of growing horticulture crops in the field.

To address the problems, by the end of 2017, the duo came up with a tech fix for solving problem of information scarcity and lack of visibility in horticulture farming and decided to launch Fasal. Fasal, meaning 'crop' in Hindi, was set up to change the farmers' problem-solving approach "from reactive to preventive". The Bengaluru-based startup officially began operations in early 2018.

The startup has built an IoT-led device with remote sensors that can be placed at farms to record crop, soil, and weather conditions. The real-time data collected by the device is harnessed using AI and ML to generate farm-specific, crop-specific, and crop-stage-specific intelligence delivered to farmers in vernacular languages through the Fasal app. Farmers can use these insights and predictions to plan their irrigation management, pest and disease management, fertiliser, fungicide and pesticide application or make any other adjustments needed to create optimal growth conditions for crops. The startup works with farmers across Maharashtra, Madhya Pradesh, Chhattisgarh, Karnataka, Andhra Pradesh, and Telangana.

The startup claims that its IoT devices cost three times lower than a typical drip irrigation system per hectare of land. Each device lasts for four to five years. "We operate on a pay-as-you-go model and the charges are customised for farmers depending on their crop requirements," say the founders.

Fasal has **raised about $1.9 million** so far, including a $1.6 million seed round led by Omnivore and Wavemaker Partners in October 2019. Fasal operates in a rapidly growing precision agriculture market that is poised to reach $6.34 billion by 2022, according to BIS Research.

Source: https://yourstory.com › Startup > Fasal

on how to increase crop yields. IBM conducted a pilot study across three states to test how its Watson Decision platform for Agriculture could assist farmers.

IBM's Watson Decision Platform for Agriculture aims to harness multiple data points and combine predictive analytics, AI, weather data and IoT sensors to provide insights on planting crops, ploughing, spraying and harvesting. Watson Decision Platform for Agriculture utilizes IBM PAIRS Geoscope, which uses geospatial data sets from satellites, drones, flights, weather models, and IoT sensors.

AI-based product interventions envisaged in the agriculture sector are:

Precision Farming and Agribots

Precision farming systems gather data at the farm-level, co-relate them with publicly available weather forecasts, and generate insights to help farmers make optimum use of seeds, water, pesticides and fertilisers. These systems also alert farmers about crop health and potential pest attacks. Agricultural robots (agribots) are used to carry out activities such as planting seeds, weeding, and harvesting.

Start-ups such as Fasal are using computer vision and AIoT sensors to collect farm-level data, which they combine with weather data to provide farmers with real-time reports detailing crop health. The data are also used to predict ideal growth conditions and resource requirements. Fasal currently operates in over six states and plans to expand across India. Similarly, Microsoft and ICRISAT, in association with the Andhra Pradesh government, have developed an application that helps farmers improve farm productivity. Using data about weather conditions and soil health, the application lets farmers know the ideal time to sow seeds. TartanSense, another start-up builds small robots for small farms. Their agribot, Brijbot, is a semi- autonomous rover that uses computer vision for de-weeding.

Farm-to-Market Supply Chains

Machine learning-based systems are being deployed to optimise agricultural supply chains, reduce produce wastage and match supply to demand. Krishihub connects farmers directly to businesses and uses ML to optimise logistics and reduce wastage. Gobasco, an online marketplace, uses real-time data analytics and algorithms to automate the supply chain and streamline it. Sensovision Systems and Occipital Tech are start-ups that use a combination of computer vision, ML and robotics to inspect produce quality and provide automated grading and sorting solutions at various points in the supply chain.

Disruptive Financial Solutions for Farmers

The computer vision and IoT sensors are being used to collect data about individual farms to improve benchmarks for insurance premiums, assess damage to speed up pay-outs, and assist government schemes with evaluations. For instance, Cropin through its service SmartRisk, generates alternate agricultural data for decision making, allowing potential lenders and insurers to carry out credit and crop risk assessments. Cropin's machine learning systems are also used to analyse the overall crop yield of a village. This has speeded up the

Box 5.4

KrishiHub, a Bengaluru based agri startup is building a technology-driven Agri ecosystem for Indian farmers through an intelligent fresh produce supply chain and an advisory platform to reduce crop wastage from farm-to-fork, increase farmer income, and improve farm productivity. The goal of KrishiHub is to improve a farmer's socio-economic conditions and increase farming productivity.

Through a demand-driven B2B platform, KrishiHub enable farmers to sell their crops directly to businesses at better price. The technology platform enable farmers to take data-driven decision in every stages of farming. Farmers across India have access to location-specific crop information, weather forecast, mandi price, agri inputs, agri news and videos and other information in regional languages through their mobile application. KrishiHub is also working to reduce crop wastage by operating an end-to-end controlled supply chain for fresh produce giving more income in the hands of farmers along with trying to improve farm productivity by providing Agri advisory to farmers across India, enabling them to make better data-driven decisions.

Source: http://yourstory.com>companies>krishihub>amp

government's decision-making under the Pradhan Mantri Fasal Bima Yojana (PMFBY), on the disbursal of crop insurance claims to farmers.

ICICI Bank has been using satellite images of farms to assess crops, and subsequently farmers' creditworthiness. IBM is developing a credit- scoring application which will help determine a farmer's average yield and profit. This will provide an estimate of the farmer's creditworthiness.

A consequence of the digital capture is the ability of financial institutions (FIs) to offer institutional credit to small holder farmers. The transactional data in combination with new age analytics based techniques can enable FIs compute risk scores for small holder farmers which in turn can help FIs price various financial products for the small holder farmers. In the current system, FIs are forced to work with joint liability groups (JLGs) due to the absence of such granular farmer level data.

Challenges in Application of AI in Agriculture

While the AI has high potential for agricultural applications, farmers still lack the use of high tech machine learning solutions. Since agriculture is affected by many external factors not under the control of farmers such as climate, soil condition, pest attack, *etc.* it can happen that what would appear a good decision in the planning stage might not prove to be the best due to the changes in external factors.

AI systems require a lot of data to train machines and arrive at making accurate predictions. It is difficult to get temporal data for large areas, though spatial data exist. Since data infrastructure requires maturity, it takes time to develop a powerful machine learning model applicable to farmers. Another disadvantage is the high cost of the AI solutions, which need to be more affordable so that technology can be accessed even at the farm level.

Box 5.5

The Ministry of Agriculture and Farmers Welfare collaborated with IBM to deploy precision agriculture solution which combines AI and weather technology to obtain and analyze farm-level insights. A pilot project to deploy IBM Watson Decision Platform for Agriculture was successfully implemented in Nanded (Maharashtra), Bhopal (MP) and Rajkot (Gujarat). The project gave insights into farm level weather forecast and village level soil moisture that were utilized to help local farmers.

IBM has also worked with Krishi Vigyan Kendra, part of Baramati, Maharashtra-based Agriculture Development Trust, to develop the 'Krushik' mobile app specifically for farmers in the state to boost overall agricultural productivity. The digital information portal, freely made available in Marathi, provides weather-based agro-advisory, agriculture-related news, crop market rates, information about government schemes and most importantly, 15 days village-level weather forecast powered by data from The Weather Company. With monsoons in India impacted in the last few years owing to changing weather patterns, the detailed and accurate weather forecast information delivered through the app has helped more than 1.5 lakh farmers make more informed decisions on their crops.

In 2019, the IBM Global High-Resolution Atmospheric Forecasting System (GRAF) was launched, the first hourly-updating commercial weather system that predicts something as small as a thunderstorm, virtually anywhere on the planet. Compared to existing models, IBM GRAF provides a nearly 200 per cent improvement in forecasting resolution for much of the globe (12 – 3 km) and can be used by farmers to better anticipate and prepare for dramatic shifts in weather.

This level of forecasting precision had been available in the U.S., Japan and a handful of Western European countries. IBM GRAF, for the first time, provides such enhanced forecasts across the globe, including Asia, Africa and South America, areas among the most vulnerable to the increasingly intense extreme weather resulting from climate change. It is the world's first operational high-resolution, hourly-updating model that covers the entire globe.

Source: http://www.zdnet.com>innovation>AI&Robotics

Data science and AI can provide accurate and actionable information which could help improve farmer productivity and incomes. Advisory solutions are expected to enable efficient consumption of water and targeted use of pesticides, making the practice more sustainable. Information about farm conditions gathered using AI and IoT can also be used to evaluate farms and provide farmers with credit and insurance.

Solutions working across the agricultural supply chain could potentially reduce wastage, match supply to demand, and improve farmer incomes. Current estimates suggest that approximately 16 per cent of produce is wasted every year.

Research has demonstrated that deep learning algorithms can effectively identify diseases across species of plants with 99 per cent accuracy. Precision agriculture systems could also reduce the amount of water used for irrigation.

Currently, the digitalisation of agriculture in India is being spearheaded by agritech startups. The introduction of AI-based technologies can further increase the dependence of farmers on large companies that operate across the agricultural value chain, selling seeds, fertilisers, and equipment, transporting produce, and also purchasing it. Most agricultural advisory interventions relate to seeds, fertilisers, and market prices. Advisory inputs, which are driven entirely by market demand, can be co-opted by companies that invest across the agricultural value chain.

5.9. Risks to Disruptive Innovation and Technology

1. Compliance and legal violations: Consumers and regulators expect companies to identify, address, and mitigate risks surrounding the privacy and protection of consumer data in the cloud. Consumers increasingly want companies to be fully transparent about where their data goes, who sees it, and what will be done with it.
2. Data breaches: Despite the best controls, data breaches happen, but some measures that have proven helpful in containing their costs have been incident response teams and the use of encryption, according to research conducted by IBM Security.
3. User Privacy: Concerns about consumer privacy reached new heights in 2018, as a number of companies suffered high-profile data breaches that compromised their customers' private information.
4. Fairness and equity: Machine learning continues to be an exciting form of disruptive technology for many applications because it offers the possibility of removing human biases from the equation when making important judgments and decisions. However, this is only effective if the dataset and model are themselves free of bias.
5. Reputational risk: Artificial intelligence left to run unchecked can represent a major and even existential risk to the business's reputation

and bottom line. Artificial Intelligence systems that are prone to errors, subject to bias, or easily hacked can expose a reputed organization to public criticism.

6. Spoofed chatbots: Impostors can create a malicious chatbot with the branding of a legitimate business and place it in an app store where unsuspecting customers of that brand seek help and download the chatbot. From there, they have a direct line to that consumer and all the sensitive information and personally identifiable information (PII) that they're willing to request to fraudulently help the client with their real query.
7. Ethical and legal concerns: As AI systems become more intelligent and gain more agency, addressing the ethical and legal issues of these disruptive innovations will be a preeminent concern.
8. Greater complexity in the IoT: Businesses need to make sure that their IoT-connected devices are safe, with no default passwords and with all security updates installed.
9. Public safety: When attackers are able to breach an IoT network that manages public infrastructure, the risks are severe. When the IoT is applied to infrastructure such as electrical grids, they must be protected with both physical security and cyber security.

5.10. Disintermediation and Uberization in Agriculture

Digital technologies today have the potential to deliver two big themes within the agriculture sector that could potentially enable small holder farmers reach their economic potential – disintermediation and uberization.

Disintermediating the Supply of Agricultural Products

Many supply chains for agricultural products have a substantial footprint in emerging economies. In the developing economies, crops are grown and harvested, aggregated, packaged, and transported for sale in domestic or international markets. Compared to supply chains that operate in developed markets, emerging market supply chains have unique features and managerial challenges. Some typical features of such chains are: (a) the absence of vertical integration, (b) a highly fragmented supply base, and (c) a weak regulatory environment. Consequently, such supply chains typically have a high degree of intermediation between the source of agricultural products and their eventual market reach (processing facilities or retailers). For example, in a typical chain, harvested crops are transported by farmers to local trading centres for sale to intermediaries (usually through auction), who in turn, may re-sell the goods through other intermediaries or the end market.

The efficiency of these supply chains are negatively impacted by their high level of intermediation, and tends to worsen over time because farmers, faced

Box 5.6

AgriTech players like FarmERP are working towards enabling a change in the agricultural world. FarmERP has been actively working for **'Uberization in Agriculture'** that aims at making farming more predictable and profitable for the smallholder farmers. This project focuses on bringing smallholder vegetable growers in Maharashtra and buyers together through a single platform.

Through its next-generation farm management system, FarmERP helps farmers and agribusinesses improve farming by the virtues of sustainability, climate-resilience, and food safety. FarmERP also helps in mapping and tracing the entire agricultural supply chain – from pre-sowing to harvesting, procurement, sales, to traceability.

The uberization project empowers vegetable growers in Maharashtra considering crucial challenges like excess rains and the brunt of COVID-19. The project covered 24 organic and conventional vegetable crops. It armed vegetable growers with end-to-end services inclusive of a package of practices, irrigation, pest and disease management, climate-smart advisory concerning nutrition, quality control, and harvesting. The mobile application was used by farmers and field officers to capture all the required data. Complete crop traceability was ensured by the labels pasted at farmgate. This allowed the buyer to know more about where their produce has come from.

The steps to vegetable cultivation were mapped by the following:

1. Geo-tagging of grower plots were administered to gain basic information, and pictures of farms and produce were put on top priority.
2. Creating a task calendar for a particular plot and notifications to growers on daily basis.
3. Climate-smart digital advisory services (DAS) in the local language are pushed as and when required, forewarning the growers on resource requirement.
4. The grower can monitor his expenses and budget for the crop cycle.
5. Agronomists intervene for specific advisories.
6. The buyer can offer purchase rates after determining the quality of produce.
7. The sale is finalized only once it's agreed by the grower, procurement, and packing.

The buyers, growers, and FPC/FPO had access to this entire process over a single digital platform. They could monitor the progress with their credentials. Meanwhile, buyers also keep track of what the farmers harvest. This allows the buyers to understand/track the quantity and quality of produce beforehand by looking at the real-time crop images that are updated regularly by the field officers. These tracking mechanics help the buyers analyze the products and pre-decide the exact quantity and best buying rates.

Source: http://www.farmerp.com

with thinning margins, are unable to make investments in their land or farming techniques to improve the yield or quality of their crops. Although contracting may be an effective mitigating approach in developed markets, contracts are often difficult to implement in emerging economies because of high supply fragmentation and weak regulatory environment. Improving supply chain efficiency in these contexts, therefore, remains a challenging problem.

Several entities (disruptors) have tried to innovate around these inefficiencies, and attempted to disintermediate the supply chain. Disruptors may be units within existing firms (*e.g.*, the eChoupal Initiative by ITC). Such disruptors aim to connect farmers more directly to the market using mobile technology. Price transparency is often a key element of disruptors' value propositions. Farmers who sell to disruptors do so at a known offered price, and do not have to face price uncertainty at the auctions. Disruptors may also offer other value-added services like transportation or packaging.

Uberization of Agriculture

Farming is an activity faced with a lot of trials and tribulation, and with little predictability of how well the crop will turn out or if it will result in a loss at the time of harvest. 'Uberization' is a term derived from the popular commute option in our fast-paced world, 'Uber'. The aim of the Uberization project carried out by FarmERP is to assist Indian farmers in initiating and formalizing market linkages, ultimately translating into a better price for their produce, analyzing data accurately, and garnering the right value of production.

All in all, the Uberization project, and multiple other such empowering activities for farmers, have enabled them to increasingly hone their independence and pave the way forward for new entrants within the stakeholder market.

Productivity and price are the two pillars on which small holder profitability hinges. Small Holder Farmers (SHFs) need to increase their yields to grow more with less. Gains in productivity can be achieved through a combination of applying good quality inputs, mechanization and knowledge. Farmers can capture better prices by selling into more lucrative and transparent markets that see participation from diversified buyer groups and also through market oriented agriculture. Coincidentally, the traditional structures and systems that deliver inputs and knowledge to the farmers in the last mile are characterized by high levels of intermediation. The agriculture input and output supply chains are intermediated due to the heterogeneous, fragmented nature of demand which is a consequence of small farm sizes and smaller lot sizes.

The agriculture markets that the small holders sell into are not only highly intermediated but are also subject to rampant abuse of free market principles in spot markets (Mandis) through cartelization. Digital commerce/e-commerce platforms supported by strong fulfilment processes can potentially disintermediate the agriculture input as well as output markets. This space has attracted the attention of a number of agritech startups like KhethiNext, Kalgudi, Agrostar *etc.* These platforms not only enable small holders buy quality inputs at lower prices directly from input manufacturers but more importantly, create a transaction history of the smallholder farmer's economic activity which has traditionally been part of the invisible rural informal economy.

5.11. Indian Digital Ecosystem of Agriculture (IDEA)

The GOI has declared that Open Digital Ecosystems (ODEs) are the next frontier for 'Digital India'. ODEs are defined as: "open and secure Digital Platforms that enable a Community of actors to unlock transformative solutions for society, based on a robust Governance framework". It identifies three layers to help bring an ODE to life digital platforms comprising technology infrastructure and solutions built on top; community comprising builders, facilitators and end users; and a third layer of governance which consists of laws and rules and the accountable institutions that uphold them.

An ecosystem approach is considered the only way forward to address the agriculture sector holistically. It is felt appropriate to draw from the national architecture principles and patterns designed by the Ministry of Electronics and IT, especially the India Digital Ecosystem Architecture (InDEA). Accordingly, the initiative is named IDEA ('InDEA Digital Ecosystem of Agriculture'). The Ministry expects that the IDEA initiative would place the farmer in the centre of the agriculture ecosystem leveraging open digital technologies.

This Ecosystem shall help the Government in effective planning towards increasing the income of farmers in particular and improving the efficiency of the Agriculture sector as a whole. As a first step in this direction, Government has already initiated building federated farmers' database that would serve as the core of the envisaged Agristack.

a-IDEA aims at fostering innovation and entrepreneurship in agriculture in India. Out of 12 start-ups being mentored by a-IDEA, four are digital start-ups that have potential to serve farmers in a better way.

Innovation Ecosystem to Connect Farmers to Disruptive Technologies

Agriculture is not immune to the changes of the digital age. Technological innovations have the ability to transform every link in the food chain, from seed to fork.

The need to embrace the opportunities these innovations offer is real. In order to feed the nearly 10 billion people with whom we will be sharing this planet by 2050, crop and livestock productivity improvements are essential. Agricultural efficiency is still relatively poor. For example, seven tonnes of feed are needed to produce just one tonne of meat. It takes 880 gallons of water to produce one gallon of milk.

Further, climate change is already requiring changes to crop management, and access to fresh water and good soil are becoming serious limitations for agriculture. There are also competing food requirements. In wealthy nations, food is a relatively small part of the household budget, and consumers are becoming prosumers, with high expectations for the standard and types of food they want. At the same time, global hunger and food scarcity are serious challenges - nearly 800 million people are undernourished. Connecting both is the global food chain: ensuring that there is transparency, traceability and trust between producers, processors and prosumers.

5.12. Challenges of Digital Agriculture

Data is one of the most common challenges we encounter. Quite simply, there is an enormous amount being produced, and only a small fraction of that is being used in a meaningful way. In many instances, agribusinesses just don't know where to start when thinking about the 'data' problem. At every level, farmers and business are producing a huge amount of data but do not have the tools to access it or derive any meaningful value. This leaves them often facing a situation where data has been poorly collected with little to no strategy, resulting in data silos that cannot be connected or utilised. It's a problem that continues to grow and will continue to get more difficult the longer it is left untackled. This is where most support is needed and why we are seeing a growing emergence of data platform businesses in the agricultural industry.

From sensors on the field to drones and weather satellites, there is highly valuable data being collected every day that could be used to deliver meaningful insight, with the right data strategies and analytic tools in place. The opportunities this could offer are wide ranging, and the potential efficiency gains are not insignificant – everything from greater visibility on risk, food traceability, and pricing transparency through to monitoring natural trends.

Agricultural cybersecurity is a rising concern because farming is becoming ever more reliant on computers and Internet access. During the last few years, the agrotechnology community, public sector, and researchers have been alerted to the problem and a significant amount of research has focused on the issue. Agriculture has adopted the use of smart technology to help meet growing food demands. This increased automation and associated connectivity increases the risk of farms being targeted by cyber-attacks. Increasing frequency of cybersecurity breaches in many industries illustrates the need for securing our food supply chain.

Another most important challenge facing digital agriculture is interoperability. With more and more OEMs (optimization of external materials) emerging with innovative agricultural IoT tools and platforms, interoperability is rapidly becoming a point of concern. The various available tools and technologies often do not follow the same technology standards/ platforms as a result of which there is lack of uniformity in the final analysis by the end users. The challenge lies in transforming the smart standalone devices and gateways to holistic, farmer-friendly platforms for attaining uniformity. In current scenario, many applications and datasets exist in silos. To get a holistic view and a standardised solution, inter-operability is critical, so that information can be exchanged among the systems and seamless integration can happen. Inter-operability addresses the open architecture of technologies and allows the software systems to interact with other systems and technologies.

Lack of data standards disables inter-operability and significantly slow down the adoption of emerging technologies. Standardization is critical to allow the production and export of data needed to support quality assessment, decision support, exchange of data. Developing a way to standardize and harmonize data is necessary, especially when working towards data interoperability among many different systems. There is need for developing standards that make inter-operability possible by providing the protocols for actual transmission of data from one computer system to another. This will support all applications thereby allowing to increased capabilities, flexibility and efficiency of operations.

Interoperability standards enable the operational processes underlying exchange and sharing of information between different systems to ensure all digital research outputs are Findable, Accessible, Interoperable and Reusable, according to the FAIR principles. Among the interoperability standards, one category focuses on the descriptions (or metadata) of digital objects. Within this category there are content standards, which opens datasets to transparent interpretation, verification and exchange.

The other important challenges are: lack of strong and reliable internet connectivity in rural areas; making sense from big data in agriculture; first mover advantage of established firms posing entry barriers for new firms; lack of scalability and configuration problems; possibility of technical failures, mostly power failures and resultant damages.

5.13. Barriers in Scaling Up Digital Innovations

Gartner has identified six barriers that organisations must overcome to transform their organization into a digital business.

1. A Change-Resisting Culture

Digital innovation can be successful only in a culture of collaboration. People must be able to work across boundaries and explore new ideas. Most organizations are stuck in a culture of change-resistant silos and hierarchies.

A digital culture should start small. Define a digital mindset, assemble a digital innovation team, and shield it from the rest of the organization to let the new culture develop. Connections between the digital innovation and core teams can then be used to scale new ideas and spread the culture.

2. Limited Sharing and Collaboration

The lack of willingness to share and collaborate is a challenge not only at the ecosystem level but also inside the organization. Issues of ownership and control of processes, information and systems make people reluctant to share their knowledge. Digital innovation with its collaborative cross-functional teams is often very different with regard to functions and hierarchies.

3. The Business Isn't Ready

Many business leaders are caught up in the hype around digital business. However, when the leader of the organisation wants to start the transformation process, it turns out that the business doesn't have the skills or resources needed.

4. The Talent Gap

Most organizations follow a traditional pattern organized into functions such as IT, sales and supply chain and largely focused on operations. Change can be slow in this kind of environment.

Digital innovation requires an organization to adopt a different approach. People, processes and technology blend to create new business models and services. Employees need new skills focused on innovation, change and creativity along with the new technologies themselves, such as AI and the IoT.

5. The Current Practices Don't Support the Talent

Having the right talent is essential, and having the right practices lets the talent work effectively. Highly structured and slow traditional processes don't work for digital. There are no tried and tested models to implement, but every organization has to find the practices that suits it best.

6. Change Isn't Easy

It's often technically challenging and expensive to make digital work. Developing platforms, changing the organizational structure, creating an ecosystem of partners- all of this costs time, resources and money.

Over the long term, enterprises should build the organizational capabilities that make change simpler and faster. To do that, they should develop a platform-based strategy that supports continuous change and design principles and then innovate on top of that platform, allowing new services to draw from the platform and its core services.

5.14. Developing Digital Trust

Technologies used for digital transformation can also be leveraged to enhance trust, when they are used to enhance transparency, reinforce ethical practices, boost data privacy, and harden security.

Technology issues like security hacks, inappropriate or illegal surveillance, misuse of personal data, spread of fake news and misinformation, algorithmic bias, and lack of transparency are regularly hitting the headlines. Nancy *et al.* (2019) have indicated the four pillars of digital trust as below:

Four Pillars of Trust

1. Transparency and accessibility- Transparency around digital business practices along with easy to understand disclosures can help build trust in an organization's intentions and its promise to deliver quality digital products and services
2. Ethics and responsibility- As technology innovations raise ethical questions by giving organizations more power, their willingness to work toward the welfare of customers can generate higher levels of credibility and trust
3. Privacy and control- Organizations that respect customer's preferences on what data to collect and how that data is handled are able to gain greater permissions to handle customer information and provide personal services.
4. Security and reliability- With heightened awareness about cyber risks and increasing reliance on smart devices, customers are increasingly changing organisations that use the latest technologies to keep products and services secure and reliable.

5.15. Reverse Innovation

In global business, the concept of "reverse innovation" indicates that some of the best ideas might flow in unexpected directions as new practices and models arise from emerging markets.

In their 2012 book, 'Reverse Innovation', Vijay Govindarajan and Chris Trimble maintain that for multinational corporations to succeed, they must move beyond the prevailing assumption that the best ideas are first developed in established markets and then exported to new ones. They indicate that some of the best ideas will come via "reverse innovation," whereby ideas first developed and adopted in emerging markets eventually gain traction in dominant regions to the benefit of all.

Reverse innovation challenges corporations by shifting the centres of control, pushing product development beyond proven ideas, and often going against conventional wisdom and best practices. However, as Govindarajan and

Trimble point out, reverse innovation is not so much about the "best practice" as it is about the "next practice" that will lead to success.

Reverse innovation acknowledges that new markets provide fertile grounds for creative thinking and problem-solving. The business world often adopts a "product out" mentality that simply looks for new markets for proven commodities. Products may be modified – often simplified – to meet lower price points, reflect regional tastes, and engage local supply chains.

Reverse Innovation is about innovating to serve the poorer populations of the world, which have third world constraints, and applying those innovations back into developed economies. There are significant lessons to be learnt for businesses in the developed world on how growth could be driven by looking further afield.

An Example of Reverse Innovation

Over 20 years ago, the manufacturer of heavy machinery Siemens was faced with a challenge when it came to selling its products in India. The impediment to increasing its market share was that the equipment needed regular maintenance. For reasons of distance, skills and logistics, this was difficult. The company charged its engineers with the task of developing a version of its equipment that required less or no maintenance, which they succeeded in doing. This approach worked so well that the company incorporated the maintenance improvements into the equipment being sold to their regular customers in the developed world, allowing them to reduce prices and increase profit and market share.

5.16. Studies of Disruptive Innovations: Implications for Extension

Curry *et al.* (2021) examined four case studies of technology adoption to explore key socio-economic factors facilitating or constraining adoption. The case studies pertained to:

1. The socio-cultural barriers to the adoption of new technologies to control cocoa pod borer in Papua New Guinea;
2. The role of land pressure on differential adoption rates of yams amongst ethnic groups in Ivory Coast;
3. Past agricultural practices and their influence on adoption of a new planting pattern and selected oil palm planting material in Cameroon;
4. Gender relations to facilitate successful adoption of a new oil palm initiative for women smallholders in Papua New Guinea.

The concept of disruptive innovation was used to assess farmers' propensity to adopt or not. It was revealed that technologies that disrupt socio-cultural and moral frameworks are less likely to be adopted. The study also indicated that communities under stress will adopt culturally disruptive technologies if they relieve that stress.

A report on scaling up Disruptive Agricultural Technologies (DAT) in Sub-Saharan Africa (SSA) prepared by the World Bank group (2019) provided the following findings:

1. There has been a spike in the number of scalable DATs in SSA. There are four key drivers of DATs in SSA: (i) low-cost and pervasive connectivity means; (ii) adaptable and more affordable tools; (iii) advances in data analytics and exchange; and (iv) increasing demand for contextualized agricultural solutions
2. Over 75 percent of scalable DATs in the database are digital. Many of the digital technologies applicable to the agriculture sector can be deployed even in low-connectivity, rural environments. Examples of DATs adapted to low-connectivity environments include portable soil testers, short message service (SMS)-based farmer education tools, interactive voice recordings, and off-line functioning platforms that can upload or receive data when a connection becomes available.
3. Thirty-two percent of the scalable DATs in the database aim to enhance agricultural productivity, 26 percent are working to improve market linkages, 23 percent are engaged in data analytics, and another 15 percent are working on financial inclusion.
4. Major barriers in scaling up DATs include limited access to finance; low technology adoption among smallholder farmers; and a lack of role models and mentoring for new entrepreneurs.
5. The biggest challenge is not the existence of solutions, but rather the enabling framework within which these DAT solutions would operate. Leading Ministries of Agriculture are seeking to systematically invest in knowledge, innovation and the incubation ecosystem for digital, women and youth entrepreneurship.

Spanaki *et al.* (2022) conducted a systematic review of AI driven agricultural technologies. They concluded that the correct definition of Agricultural Technology (AgriTech) is strongly associated with AI applications and referred to the progression from conventional farming to smart farming, characterised by the exploitation of disruptive technologies (*e.g.* multi source data, sensors on farm equipment and plants, satellite images, monitoring of input use, *etc.*) for precision farming.

The study identified three categories (physical, cyber and cyber-physical) according to specific technological applications. The physical AgriTech application types are defined as the disruptive technologies for agricultural operations which can replace not only human labour tasks, but also present physical features as the 'hardware' of AgriTech. The cyber aspects are applications which are mostly platform software related. The cyber-physical application is the combination of both cyber and physical, which include the hardware and software for data analysis and prescriptive tailored decision making and recommendations.

The research studies on AgriTech physical aspects mostly consider land operations. The AgriTech cyber aspects can be categorised into three forms-analytics, virtual/stimulation and algorithmic based on the different types of tools and techniques applied. There are numerous studies about analytic applications on land for plants that use a wide range of AgriTech tools. There are also many studies related to AgriTech algorithm applications for land and plants. Some of the studies about AgriTech analytical cyber-physical applications include:

- ✰ A suboptimal path for agritech mobile robots combining neural network methods and genetic algorithms.
- ✰ IoT enabled platform for research in precision agriculture and ecological monitoring domains
- ✰ Decision making system for intelligent chemical control of pests

The review concluded that the role of Operations Management is vital to bridge the research gap between 'Agricultural Technology' and 'Sustained Agricultural Operations' and equip the future generation of farmer agripreneurs.

REFERENCES

Brower, J L and Christensen Clayton (1995) Disruptive Technology: Catching the Wave, Harvard Business Review, 73 (1), Jan-Feb 1995, p 43-53

Christensen Clayton (1997) The Innovator's Dilemma, Harvard Business Review Press

Christensen Clayton and Michael E. Raynor (2003) The Innovator's Solution: Creating and Sustaining Successful Growth, Harvard Business Review Press

Christensen Clayton, Heiner B, Rudy, R and Thomas, M S (2006) Disruptive Innovation for Social Change, Harvard Business Review, Dec 2006

Christensen Clayton, Jerome Grossman and Jason Hwang (2009) The Innovator's Prescription: A Disruptive Solution for Health Care, McGraw Hill Education

Curry N George *et al.* (2021) Disruptive innovation in Agriculture: Socio-cultural factors in technology adoption in the developing world, Journal of Rural Studies, Vol 88, Dec 202, p 422-431

Department of Agriculture, Co-operation and FW, GOI (2021) Consultation Paper on IDEA, June, 2021

Hunter, Evan (2016) Reverse innovation: Moving beyond Best practices, Insights Journal for Global Theological Education, Vol 1, No.2 p 9-13

Jeff Dyer, Hal Gregersen and Christensen Clayton (2011) The Innovator's DNA: Mastering the Five Skills of Disruptive Innovators, Harvard Business Review Press

Nancy Albinson, Sam Balaji, Yang Chu (2019) Building digital trust: Technology can lead the way, Deloitte insights, September 2019

Spanaki Konstantina, Uthayasankar Sivarajah, Masoud Fakhimi, Stella Despoud, Zahir Irani (2022) Disruptive technologies in agricultural operations: A systematic review of AI driven AgriTech research, Annals of Operations Research (2022) 308:491–524

https://doi.org/10.1007/s10479-020-03922-z

Tandem Research (2020) AI on the Ground: A Snapshot of AI use in India, Tandem Research Report, Goa, October 2020

World Bank (2019) Scaling up Disruptive Agricultural Technologies in Africa, Executive Summary August 2019

https://olc.worldbank.org › system › files

Chapter 6

Innovation and Entrepreneurship in Agriculture

Innovation and entrepreneurship are recognized as key building blocks of competitive and dynamic economies. Creativity and innovation have become the vital forces to enhance the value of entrepreneurship.

6.1. Creativity in Entrepreneurship

In a world characterized by volatility, uncertainty, complexity, and ambiguity, leaders require robust innovation skills. Thinking flexibly and developing an entrepreneurial mindset are critical to thriving in uncertain business environments. Learning to change the rules of the game by creating innovative value propositions and discovering new market positions for sustained competitive advantage are important in this respect.

Entrepreneurial creativity is defined as the generation and implementation of novel, suitable ideas to establish new venture. The process of creating new venture is inherently to be dynamic and versatile. The phrase 'thinking outside the box' is often used for creative problem solution and the expression finds application in psychology, business and marketing. Entrepreneur need to think out of box to enhance the entrepreneurial qualities.

Creativity is the intellectual act to create new ideas while innovation is the action taken to transform the new ideas into a result in the form of a product. In a competitive market, creative entrepreneurs must be able to value add the speciality in their product and services, which could attract customers.

6.2. Sources of Innovation in Entrepreneurship

Peter Drucker's seven sources of innovation provides a purposeful way to search for opportunities as well as mitigate risks in entrepreneurship. They are:

1. **Unexpected success or failure**: Understanding the reasons for the unexpected success or failure of a product generates opportunities to innovate. When something unexpected happens, there is a tendency to overcome it, which is a learned behaviour. If 'The Unexpected' is fully analysed, there is every likelihood to find an opportunity to innovate.
2. **Incongruity between what actually happens and what was supposed to happen:** If things are not happening as they should, there is scope to innovate. Many people will be following a particular practice and operating in a way that is the status quo, instead of asking "is there a better way to do a particular function". This is an indicator there may be a case of incongruity, because they are bottom line if something conflicts with the status quo, they are not inclined to change it.
3. **Deficiencies in a process, that are taken for granted:** If a process is inefficient or suffers from a big gap, there is scope to innovate. One can always analyse a process and find concrete ways to improve it. Amazon is a prominent leader in this respect. If one can take time to study a process, the possibilities to innovate emerges, especially as technology improves.
4. **Changes in industry or market structure that catch everyone by surprise:** The emergence of new, fast-growing segments provides scope for innovation. Innovators can serve the needs of these segments. A change in an industry or market opens up ample opportunity to innovate. Cloud technology is probably the most appropriate example for this. Since internet services and speeds have significantly improved, the door for the cloud computing market has opened.
5. **Demographic changes:** Demographic changes result in new wants and new lifestyles that call for new products. Innovation focussed on demographics can provide insights on the design and performance of products suited to various profiles. Demographics can impact on the engineering aspects of a product.
6. **Changes in Perception:** By changing the common perception of people, new needs can be created. For example, capitalizing on people's concern for health and fitness, a booming industry has emerged for exercise and jogging equipment. Though Amazon is an online retailer,

seeing an opportunity in cloud computing, Amazon had built elaborate data-centres to run their business. They changed the perception of what business they were in and started the Amazon Web Services, as part of their business to capitalize on the rapidly expanding cloud business.

7. **Changes brought about by New Knowledge:** New knowledge can be used to develop innovative products. Innovations of this sort usually combine many sorts of knowledge. There is usually a gap between the emergence of new knowledge and its conversion into usable technology and another gap before the product is launched in the market. Conceptualising a new idea and working on that idea to develop an entirely new market is the essence of innovation. This is where the Bill Gates, Mark Zuckerbergs, and Steve Jobs have succeeded.

6.3. Technology Entrepreneurship

Successful entrepreneurs realize that new ventures usually do not fail due to lack of planning, but they fail due to lack of customers. Seasoned entrepreneurs know they must become exceedingly familiar with their market and customers and understand their wants and needs in their terms. In other words, smart entrepreneurs identify, study, and become deeply familiar with customer problems before designing and commercializing a solution.

The lean start-up is an approach to venture development that emerged from the venture development experience of Eric Ries, who has captured this experience in his highly popular book used by entrepreneurs around the world.

After the failure of the first start up created by Eric Ries and colleagues, they developed a new approach where they released features to customers in an iterative manner. They released what they called a "minimal viable product" (MVP) and tested it with real customers. Releasing MVPs in this iterative manner enabled the start-up to learn what customers wanted before they invested precious resources. In other words, they developed a strategy for starting ventures that mimics the lean approach to manufacturing perfected by Toyota. The approach is designed to minimize costs and maximize learning opportunities.

The recognition that entrepreneurs begin their quest to launch a new venture by taking stock of who they are, what they know, who they know, and the resources they control has been condensed into what is referred to as 'effectuation'. In fact, it has a powerful and important operational sense that all aspiring and practicing entrepreneurs should understand. One of the more important discoveries made by effectuation researchers is the difference between "causal logic" (selecting resources to achieve a predetermined goal) and "effectual logic" (imagining a possible new end using a given set of resources).

Box 6.1

Dr. Saras Sarasvathy proposed the theory of effectuation in the early 2000s after studying a sample of expert entrepreneurs with diverse backgrounds. Effectuation theory is often considered a process theory because it explains the process that entrepreneurs use to create new ventures.

Effectuation logic contrasts with what Sarasvathy calls "causation theories" of entrepreneurship, where it is proposed that entrepreneurs start with a goal and then acquire the resources needed to achieve the goal, in a linear fashion. Each resource acquisition is a step toward the goal.

In contrast, effectuation logic involves evaluating resources available to use today and then deriving goals out of what can be made from the recombination of those resources. Thus, entrepreneurs don't just recombine resources to meet goals, they accept floating goals within a set of limits and allow the resources that are available to guide the evolution of their strategies.

Elaborating on the theory, Sarasvathy suggested that effectuation involves five core principles.

1. The bird in the hand principle: A bird in the hand beats two in the bush. This refers to maximizing the use of what an entrepreneur knows (their background and experience), who they know (family and peer groups), and aligning options based on who they are (what are the entrepreneur's abilities).
2. The affordable loss principle: Only take on affordable losses. Don't obsess about windfall profits, but do try to minimize potential losses. This involves taking low probability bets, but only with a small investment of resources lost with each failure.
3. The crazy quilts principle: Make crazy quilts. Weave potential deals with potential partners until something sticks. Many iterations are often required.
4. The lemonade principle: Make lemonade means to see potential in depressed or under-utilized resources. This is similar to the alertness principle.
5. The pilot in the plane principle: Focus on today, not next year.

Source: entrepreneurshiptheories.blogspot.com/2017/08/sarasvathy-effectivnation-theory.html

Technology entrepreneurship is an investment in a project that assembles and deploys specialized individuals and heterogeneous assets intricately related to advances in scientific and technological knowledge for the purpose of creating and capturing value for a firm. The definition of technology entrepreneurship is based on four elements:

1. **Ultimate outcomes:** Value creation and capture are identified as two outcomes of technology entrepreneurship because the sources that create value and the sources that capture value may not be the same over the long run.
2. **Target of the ultimate outcomes:** The firm is identified as the target organization for which value is created and captured.
3. **Mechanism used to deliver the ultimate outcomes:** Investment in a project is the mechanism mobilized to create and capture value. A project is a stock of resources (*i.e.*, specialized individuals and heterogeneous assets) committed to deliver the two ultimate outcome types for a period of time.
4. **Interdependence of this mechanism with scientific and technological advances:** The individuals involved in a project influence and are influenced by advances in relevant scientific and technology knowledge. The project exploits or explores scientific and technology knowledge. External and internal individuals and organizations co-produce the project's outputs.

The primary function of technology entrepreneurship is to assemble a combination of specialized individuals and heterogeneous assets in order to create and capture value for the firm through collaborative exploration and experimentation. The combination, some of the assets, or the assets' attributes may be unique and novel. The initial combination may change over time.

There are two primary steps of technology entrepreneurship, the characteristics of each raising different business issues:

a) Technology developers- those who develop a unique technology capable of diving a new business,

b) Technology users- those who see a new technology development and understand how it can be applied to meet a market demand

6.4. Innovation and Technology

Technically, innovation is larger than research and development as it includes three definite parts: discovery, incubation, and acceleration. Predominantly, R&D is only a fragment of the first stage.

Technology is simply a driver of successful innovation. It operates within an ecosystem of individuals and non-profit, government, and private organizations focused on the social and economic use of new products, processes, and organization forms. Technology alone will not produce enough nor sustained success in any sector, including agriculture.

Most innovation failures stem from failing to recognize that innovation is both an economic and social process rather than simply a technical one. Many innovation endeavours fail despite funding availability, talented people, strong technology, and sincere intentions.

Failure to address disruptive change and implement business model innovations has led to the demise of many successful companies and organizations, in both developed and developing countries.

6.5. Technology Ecosystem Model

Technology ecosystem is used to emphasize the inherently organic nature of technology development and innovation is often absent in standard forecasting and analytical methods.

The conceptual model builds on hierarchical models of technology evolution, most notably by Rosenkopf and Nerkar (2001). In this model three levels of the hierarchy are typically considered, component level technologies combine to form product level technologies and products are then combined to form a system of use. Coevolution of technologies in this model occurs both within levels and across levels in the hierarchy.

For example, an ecosystem of personal computer technologies would include the PCs themselves, as well as the components (hard disk drives, processors),

Box 6.2

Cassava peels discarded during processing make up 20 per cent of the whole root. Cassava had earned a bad name as an environmental polluter with the mountains of waste around processing centres. To create a business opportunity out of this undesirable bye-product, the International Institute of Tropical Agriculture has developed high quality cassava peel (HQCP) feed ingredients from wet peels. This innovation enables rapid water removal and accelerates the elimination of hydrocyanide.

A study was conducted by the Department of Animal Sciences, Obafemi Awolowo University, Ile-Ife, Nigeria, which investigated the growth response and economic performance of using HQCP meal as a replacement for maize in the diet of African catfish. The growth parameters, value of fish produced and incidence of costs in this study suggested that high quality cassava peel can successfully replace maize in the practical diets of A. catfish.

The huge value creation of this high impact innovation provides an alternative source of feedstuff, protects the environment and provides new income sources to smallholder producing cassava. It has been supported by the CGIAR Research Programme (CRP) on Root Tubers and Bananas (RTB) and it leverages the expertise of several private and public partners in Nigeria.

Source: Bulletin of Animal Health and Production in Africa 2019 Vol.67 No.1 pp.67-77.

competing technologies (laptops), and peripheral or collaborating technologies (printers, software, scanners). Additionally, it is apparent that component-level technologies are necessary for the development of PC technologies. Similarly, peripheral technologies such as software and printers require the existence of the PC technologies in order to add value to the system of use. The hierarchal nature of technologies within a population leads to the identification of specific roles that technologies can play within the ecosystem. By acting through these roles, classes of technologies can influence the evolution and development of each other. Specifically, three roles that technologies can play within an ecosystem could be identified: component, product and application, and support and infrastructure.

The component role describes technologies when they are used as components in more complex technologies. For example, there are several technologies that act as components for the personal computer: microprocessors, RAM chips, hard disk drives, *etc.*

The product and application role describes technologies when they are built up from a set of components, and are designed to perform a specific set of functions or satisfy a specific set of needs. When acting in this role, the technology is associated with a specific application or use.

The support and infrastructure role describes technologies when they work in conjunction or collaboration with (or as a peripheral to) other technologies. The distinction between the component role and the support and infrastructure role is that components are necessary for the design and are part of the physical structure of another more complex technology, while support and infrastructure technologies simply work in combination with other technologies.

6.6. Entrepreneurship and Social Capital

Social capital is a concept that refers to the value derived from connections within and between social networks. These relationships are built on mutual belonging and trust. Benefits from social capital are derived through the goodwill that members of networks feel toward each other.

The question of whether entrepreneurs are more likely to succeed if they have access to greater resources and information has been extensively studied. Considering the social capital field, this has meant looking at the connections that entrepreneurs have to networks of people who have been in business, are party to knowledge on a sector that might be useful, or can provide connections to financiers. Studies have shown that the personal networks that entrepreneurs can draw on add 'external resources' to the 'internal resources' of ambition and ideas that they already have.

One prominent theory divides the relationships that provide these extra resources into three types: structural, relational and cognitive. The structural aspect is essentially the ability that entrepreneurs have to access things through a network. The more access and the bigger or more extensive the network, the better it is. The relational aspect refers to the quality of the relationships in that network- how much mutual respect, trust and friendliness exists between the people involved. Finally, these networks are changed by the extent of shared norms and ideas- the cognitive aspect. A good example of this is the tolerance of entrepreneurial spirit and the risks attached to it. The more it is accepted, the more a network is likely to be open to failure.

Dana and Light (2012) had indicated that social capital sometimes enhances and sometimes obstructs entrepreneurship. Social capital encourages entrepreneurship when people can access essential resources via social networks, thus conserving their time and money. Membership in multiple social networks also enhances people's mental capability to perceive opportunity, additionally encouraging their entrepreneurship. Social capital's obstruction occurs when closed social networks exclude prospective entrepreneurs from essential resources, reward safety and mediocrities rather than risk and entrepreneurship, or impose mental conformity upon network participants.

Some studies have found that these factors work together in certain ways and that some of them might benefit particular types of entrepreneurs more than others. For example, there is evidence to suggest that budding entrepreneurs

Box 6.3

The Indus Entrepreneurs (TiE) is a global not-for-profit organization with mission for nurturing startups. TiE was founded in 1992 with headquarters in US by a group of successful entrepreneurs and is currently the world's largest entrepreneurial organization. With a global reach and a local focus, the heart of TiE efforts lies in its five foundational programmes- Mentoring, Networking, Education, Funding, and Incubation.

TiE Chapters around the world has become a vibrant platform for entrepreneurs, professionals, industry leaders, investors to interact with one another and forge long lasting relationships. They provide a platform to support an ecosystem for entrepreneurship by facilitating linkages between entrepreneurs, investors, mentors and professionals. TiE aims to link entrepreneurs' needs with the sectoral background and skills of the mentor. An example of services provided by mentors can include anything from market research, business strategy, human resourcing, to more hands-on roles such as active management, seed funding and in some cases, sharing overheads such as administrative support and office space. Mentors also assist young entrepreneurs in navigating the contextual challenges in India including understanding how to structure delivery models that can account for the cultural disparities between India's many villages.

Source: https://tie.org

need to first start using their ties, then to start contributing to the norms and views of their network, and by doing so develop bonds of trust with people who can provide valuable support. Furthermore, high-tech entrepreneurs have been found to gain substantially from relationships and networks that have strong, trusting connections which give them the access to information and knowledge that they need.

6.7. Business Innovations in Agriculture

Natural Resources Institute (NRI) has developed the Social Venture Impact Metric (SVIM) to assess three types of impact resulting from business innovation. NRI is a specialist research, development and education organisation of the University of Greenwich, United Kingdom with a focus on food, agriculture, environment and sustainable livelihoods.

Business innovations can have a profound impact on low income populations, referred to as Base of Pyramid (BoP) population. These innovations may be designed to address development problems faced by BoP population. They may take the form of a product (eg: low cost solar drier), a technology (eg: climate smart technology for dairy farmers) or an application (eg: IT app that connects farmers to markets), *etc.* Ultimately, business innovation has multiplier effects.

In SVIM, three levels of impact are measured:

1. Primary, the impact of innovation on low income/BoP populations- Economic well-being, capability well-being and relationship well-being are focused in the primary impact.
2. Secondary, the impact of innovation partnerships, which could be at the enterprise level, and measured through assessing both financial and environmental performance of the business – Enterprise performance
3. Tertiary impact, which is realized when the innovation is able to catalyse market-level change. This means achieving greater outreach and a critical scale in the market and/or accelerating the development of a market that reaches an underserved or disadvantaged population. It also means understanding the 'market-disruption' (both positive and negative) potential of the innovation- Market system development

Tertiary impacts bring much more systemic change, leading larger numbers of businesses and policy programmes to learn from the innovation and join the bandwagon. Positive market system development can usher in local economic development and social domino effects. Market-level systemic impact is possible, even though much harder to achieve and to measure.

This metric provided a method and tools geared specifically towards agribusiness and other sectors.

Box 6.4

The Technology Innovation Institute (TII) is part of Abu Dhabi Government's Advanced Technology Research Council, which oversees technology research in the emirate. As a disruptor in science, TII is setting new standards and serve as a catalyst for change. TII aims to become a leading global research Centre dedicated to pushing the frontiers of knowledge. The teams of scientists, researchers and engineers work in an open, flexible and agile environment to deliver discovery science and transformative technologies. The rigorous discovery and inquiry-based approach helps to forge new and disruptive breakthroughs in:

- ☆ Quantum computing
- ☆ Autonomous robotics
- ☆ Cryptography
- ☆ Advanced materials
- ☆ Digital security
- ☆ Directed energy
- ☆ Secure systems

Source: https://www.tii.ae

6.8. Developing Entrepreneurship in Agriculture

Entrepreneurship is a concept that encompasses transforming an idea or vision into a "new business or new venture creation, or the expansion of an existing business, by an individual, a team of individuals, or an established business". It plays a vital role in bringing in new technologies, innovations and economic changes. Activities like diversification, value addition, precision farming, high-tech agriculture, agripreneurship, global marketing, organic farming, *etc.* are gradually getting due attention of people involved in redefining agriculture.

Small farmers possess technical and managerial skills but lack entrepreneurial drive to take calculated risks to overcome agribusiness challenges related to accessing remunerative markets, timely finance, low bargaining power, regulations, appropriate technologies and information, and access to BDS providers.

Farmers need improved and innovative production systems to increase their production line and also to address the various challenges faced by them. Agriculture today faces many challenges, and many of these directly or indirectly lead to changing markets, which create both opportunities and risks for farmers. With a growing recognition of the important role of smallholder agriculture for economic growth and rural development in many countries, market-oriented agriculture has emerged an important agenda. Agripreneurship is key in this regard.

Agripreneurship refers to entrepreneurship in the field of agriculture and allied activities. The major agripreneurship opportunities exist in agricultural inputs, farming processes, and technology, value chain, agriculture output processing, and in allied sectors like dairy development, poultry, fisheries and so on. Activities like diversification, precision farming, high-tech agriculture, organic farming, *etc.* are also adding sustainable value.

In agripreneurship, the most important issue that needs to be understood is that the business needs to have a much more enhanced commercial activity that involves trade and trading in every form. With the changing market dynamics, many choices are available to the consumers. The agricultural producers and especially the agricultural companies have to adapt increasingly to the market demand, changing consumer habits, enhanced environmental regulations, new requirements for product quality, chain management, food safety and sustainability.

With government incentive and financial assistance, the scope and potential of agripreneurial opportunities is increasing. There are very promising opportunities for agripreneurial process in the areas such as bio-pesticides, bio-fertilizers, vermicomposting, soil testing, agro service, post-harvest opportunities in the areas of value chain, output processing, marketing, cold chain, logistics, *etc.*

Agriculture is a business like any other venture and it should be treated the same way other businesses are treated. One way to treat agriculture as a business is to get the private sector more involved, since government alone cannot create agricultural transformation. The government can intervene by putting right policies and regulations in place, by creating strong institutions, and by building sufficient infrastructure. But the real transformation has to be led by the private sector since farmers need access to finance, inputs, information, or markets, where government cannot do much.

The entrepreneur can create new combinations of production factors such as new methods of production, new products, new markets, finds new sources of supply and new organizational forms; or as a person who is willing to take risks; or a person who, by exploiting market opportunities, eliminates disequilibrium between aggregate supply and aggregate demand, or as one who owns and operates a business.

Agripreneurs as Agribusiness Providers

A business service provider manages and operates standardized business processes on behalf of its customers, delivering its service across a network to

Box 6.5

United Nations Industrial Development Organisation (UNIDO) aims to promote sustainable, inclusive business opportunities for the rural poor through agri-business and agro-value chain development. UNIDO's technical cooperation activities in this regard focus on adding value to agricultural commodities including non-food sectors at various points of the chain of economic transactions that links input providers, farmers, traders, processors, logistic providers, distributors and retailers.

Through its technical assistance, UNIDO links resources and markets in the agribusiness value chains and strengthens forward and backward industrial linkages in order to leg up the economic transformation of countries, improve employment and income opportunities, and reinforce sustainable livelihoods.

Agro-industrial activities benefit a number of groups, including poor and marginalized rural populations, urban agro-industries and communities facing human security challenges or requiring urgent supplies of agricultural equipment and the rehabilitation of food industries. Technical cooperation and capacity-building services are provided to agro-based and agro-related businesses and industries, inter alia, in the food, leather, textiles, wood and agricultural equipment sectors.

UNIDO promotes investment in agribusiness and value chain development; builds partnerships and linkages with strategic financing institutions; organizes various global forums and expert group meetings in related fields; and publishes specialized training manuals, guides and electronic media.

To carry out its mandate in this area, UNIDO mobilizes expert services such as cluster development, conformity with quality and standards, rural energy, environmental management and cleaner production.

Source: http://www.unido.org>department-agribusiness

multiple customers. Thus a Agri business provider can extend various services by taking different roles such as:

- ☆ Business mentor: Farmers generally know how to produce crops and raise livestock, what they want to learn is how to improve their farming business skills from established business mentors, who can advise them on how to make investments to improve the existing enterprises.
- ☆ Marketing expert: Expert who can work with farmers to identify new markets to explore, unique products to sell, *etc.*
- ☆ Production and Technology expert: Specialist who can provide information on how to improve productivity, without compromising quality, through state of the art technological options.

- Postharvest handling expert: Specialist who support value addition, packing, branding, *etc.*
- Financial services expert: Provide advice on the most appropriate financing method to be selected.
- Value chain specialist: As one who can work on ways to strengthen investments across a value chain.

Risk Management by Agripreneurs

Successful entrepreneurship involves taking risks. Taking risks in entrepreneurship involves careful planning and hard work. Farmers make decisions every day that affect farming operations. Many of the factors that affect the decisions they make cannot be predicted with complete accuracy; this is risk. Farmers as well as agripreneurs need to understand risk and have risk management skills to better anticipate problems and reduce consequences. Agricultural risks are classified into different categories as below:

Production and Technical Risks

These are the major threats that always remain as concerns among the farmers. The quantity and quality of commodities produced is directly impacted from weather, disease, pests, and other factors like market preference, government policies, availability of inputs including farm machinery, labour strike/lockdown, *etc.*

Financial Risks

Financial risk occurs when money is borrowed to finance the farm business. This risk can be caused by uncertainty about future interest rates, a lender's willingness and ability to continue to provide funds when needed, and the ability of the farmer to generate the income necessary for loan repayment. Many farmers due to their poor financial condition may not be able to access high quality inputs or labour at the right time. Farmers often sell their produce to traders who have provided them with credit for their inputs and are unable to negotiate good prices for their produce and this traps them in a cycle of financial debt and low income. Production risks often result in financial risks. Possible cash flow problems may come from the fact that agricultural production cycles are characterised by long periods of low sales income. In addition, problems may arise from limited access to financial sources.

Market-Related Risks

Factors affecting financial risk include supply, demand and cost of production. Changes in prices are beyond the control of any individual farmer. The price of farm products is affected by the supply of a product, demand for the product, and the cost of production. Sometimes price movements follow seasonal or cyclical trends that can be predicted. Many times, however, supply or demand will change unexpectedly and, in turn, affect the market price.

Smallholder farmers often face difficult choices, particularly about which markets to serve. When advising farmers on market opportunities, it is important that they are matched with appropriate types of markets and risk is a key factor in making decisions on the suitable market.

Institutional Risks

Institutional risk refers to unpredictable changes in the provision of services from institutions that support farming. The agricultural sector is characterised by strong government interventions and regulations, *e.g.* subsidies and price control. However, changes in government regulation may affect agricultural producers unexpectedly and may, for example, lead to changes in import/export conditions. Export bans are a commonly used government policy when production falls within their own borders, or in a neighbouring country. This type of decision can have drastic effects in the marketplace and may lead to considerable loss of income for farmers. The farmer co-operatives that do not manage their resources carefully can also have dire consequences on farmers who have invested in the co-operatives for either inputs or market access.

Human and Personal Risks

Human risk refers to the risks to the farm business caused by illness or death and the personal situation of the farm family. Accidents, illness and death can disrupt farm performance. In many states labour migration is a common occurrence. Migration can cause labour shortages for the farm. Political and social unrest can also limit labour availability.

Agri Clinics and Agri Business Centres (AC-ABC) Scheme

AC-ABC programme is launched by the Ministry of Agriculture and Farmers' Welfare, GOI in association with NABARD and implemented by MANAGE. The scheme provides extension services to farmers through setting up of economically viable self-employment ventures. It aims to tap the expertise available in the large pool of Agriculture graduates, who can set up Agri Clinic or Agri Business Centre and offer professional extension service to farmers on a payment basis. These Centres established by trained professionals are utilized to support agricultural extension services in the state.

The MANAGE provides agripreneurship training for starting Agri clinics or Agri Business centres through identified Nodal Training Institutions in various states, which will also offer handholding support for one year. NABARD offers credit support to Agri-Clinics through Scheduled Banks. MANAGE is also providing start-up training to graduates in Agriculture, or any subject allied to Agriculture like Horticulture, Sericulture, Veterinary Sciences, Forestry, Dairy, Poultry Farming, and Fisheries. On completing the training, the trainees can apply for special start-up loans for venture. AC and ABC Incubation Centre is established in MANAGE.

Knowledge Partners (KPs) as Centres of Excellence

The Department of Agriculture and Farmers Welfare, GoI has selected five Knowledge Partners (KPs) as Centres of Excellence under the Scheme for Innovation and Agri-Entrepreneurship Development

- ✰ National Institute of Agricultural Extension Management (MANAGE), Hyderabad,
- ✰ National Institute of Agricultural Marketing (NIAM), Jaipur,
- ✰ Indian Agricultural Research Institute (IARI), Pusa, New Delhi,
- ✰ University of Agriculture Science, Dharwad, Karnataka and
- ✰ Assam Agriculture University, Jorhat, Assam

Agribusiness Incubators (R-ABIs)

The Agri-Business Incubation (ABI) Programme at ICRISAT, launched in 2003, is an initiative of the ICRISAT in partnership with India's Department of Science and Technology (DST). ABI promotes agricultural technologies developed by ICRISAT, other R&D centres of excellence, universities, and other institutions, separately and jointly. These incubators have stakeholders in government agencies, financial institutions, and venture capital.

ABI represents a new resource to promote enterprise development in agriculture and facilitate business among entrepreneurs and technology developers. The pillars for high-performance incubation are R&D, business planning, business development, and access to capital

Aakruthi Agricultural Associates of India (AAI) was launched in 2004 as a commercial alternative to government agricultural extension services in Anantapur district of AP. AAI joined forces with Agri Business Incubation (ABI) in 2005 and 2006 and emerged as the second venture to graduate from the incubator.

Through the incubator, AAI developed a successful business model based on partnership with seed providers of ICRISAT, national and state agricultural research centres and farmer franchisees.

Business incubators play a key role in providing assistance to nascent entrepreneurs, particularly in the initial stages of the firm's life cycle. 22 Agri Business Incubators were established by ICAR through its WB funded NAIP-10 ABIs in 2008-09 and 12 in 2013-14. These ABIs have given a boost to the technology based entrepreneurship in agriculture. Within a span of 5 years, more than 100 technologies have been commercialized.

24 RKVY-RAFTAAR Agribusiness Incubators (R-ABIs) from across the country have also been appointed for the efficient implementation of the scheme. The funds will be released in instalments, and the selected startups will be trained for two months at various agribusiness incubation centres spread across India.

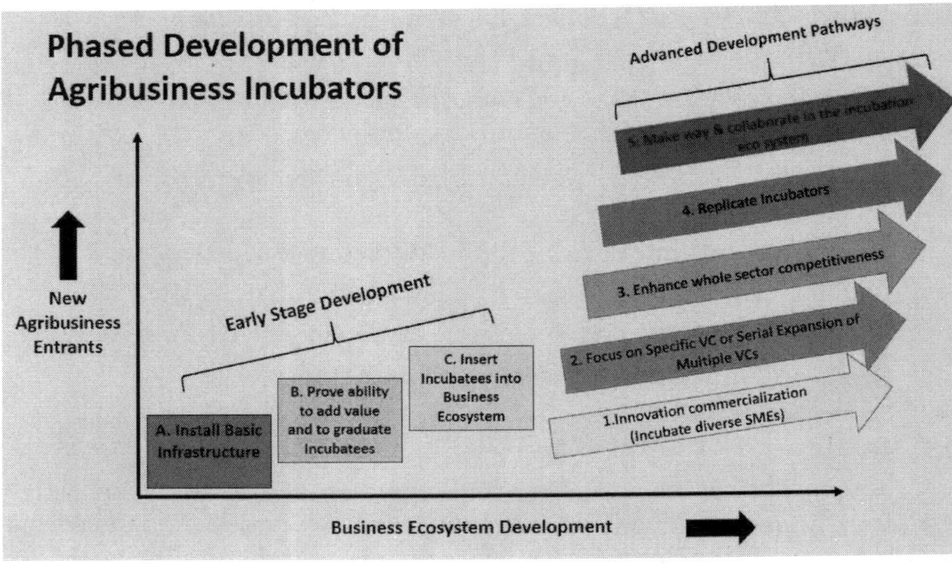

Figure 7: Phased Development of ABIs.

Based on a rigorous process of selection through various stages and training of two months, the final list of start-ups that are to be funded through grants-in-aid will be finalized. Training on technical, finance, intellectual property, statutory compliance issues, *etc.* will be provided. Mentoring of start-ups through monitoring of milestones and timelines is part of the programme.

Scheme for Innovation and Agri-Entrepreneurship Development

Ministry of Agriculture and Farmers Welfare, Government of India launched a Scheme titled 'Innovation and Agri-Entrepreneurship Development" under the revamped Rashtriya Krishi Vikas Yojana (RKVY) to promote Agri-entrepreneurship and agri business by providing financial support and nurturing the incubation ecosystem. Under the scheme, grants are provided for start-ups to build on their business ideas and for those who are already in business and want to scale up, and the release of funds is based on the achievement of milestones.

The Nodal Division of Rashtriya Krishi Vikas Yojana- RAFTAAR (Remunerative Approaches for Agriculture and Allied Sectors Rejuvenation) will be responsible for governance and overall management of the scheme. The key objective of the scheme is to contribute directly and indirectly to enhance the income of farmers by providing employment opportunities to young entrepreneurs. The details of the scheme are given in **Annexure-5**.

Intellectual Property (IP) Protection Regulations

Intellectual Property (IP) is the domain of law which protects the freedom of those creating initial intellectual and novel works. It includes everything from innovations and inventions. World Intellectual Property Organization (WIPO) of the United Nations seeks to develop a balanced and accessible International Intellectual Property system, which rewards creativity, stimulates innovations, and contributes to economic development while safeguarding public interests.

Some of the legal instruments passed by the Indian Parliament as part of compliance process to the IP protection include The Patents Act, 1970 (39 of 1970), The Patents (Amendment) Act, 1999 (17 of 1999), The Patents (Amendment) Act 2002 (38 of 2002), The Patents (Amendment) Act 2005 (15 of 2005), The Geographical Indications of Goods (Registration and Protection) Act, 1999, Patents Designs and Trade Marks, 2016 and The Protection of Plant Varieties and Farmers Rights Act, 2001 (PPV FR Act).

The details are given in **Annexure-6**.

6.9. Agri Startups

A startup is a young company founded by one or more entrepreneurs to develop a unique product or service and bring it to market. Startups are common in industries. Transformation of agriculture to agribusiness is one of the important strategies where enterprising farmers practice profitable agriculture. Modern techniques and methods can elevate agriculture to the next level and ease the burden on farmers. This creates a huge scope for Agriculture Startups in the country.

Acording to Eric Ries (2011), the creator of the Lean Startup Methodology, "A startup is a human institution designed to create a new product or service under conditions of extreme uncertainty."

The website "startup commons" defines startup as a venture that is initiated by its founders around an idea or a problem with a potential for significant business opportunity and impact. It is a team of entrepreneurial talent developing new innovations, in identifiable and investable form, in progress to validate and capture the value of the created innovation- with ambition to grow fast with scalable business model for maximum impact (www.startupcommons.org).

Startups can provide missing links in the agri value chain and deliver efficient products, technologies and services to the farmers on one hand and the consumers on the other. From ICT apps to farm automation and from weather forecasting to use of drone and from inputs retailing and custom hiring to online vegetable marketing, and from poultry and dairy ventures to smart agriculture and from protected cultivation to innovative food processing and packaging, there are over 250 Agri-Tech startups in India across the value chain, leveraging the use of technology and innovation in business models to impact the large agricultural sector in India.

India's digital ecosystem is witnessing healthy signs such as affordability and availability of high-speed internet and maturing digital content ecosystem. The confluence of these factors presents an exciting opportunity for innovation in agricultural ecosystem, wherein market players, including start-ups, can leverage next generation technology such as data digitization and data platforms, data analytics, Artificial Intelligence, Machine Learning, the Internet of Things and Software as a Service (SaaS) to disrupt the status quo.

Adoption of technology in agriculture (agritech) helps in solving several pain-points across the spectrum of traditional agriculture value chain and presents a market potential, as indicated by Pahwa (2020).

Pain-point	*Potential of Agri tech Segment*
Volatility in public prices, Sub optimal impact selection	Market linkages- Farm inputs
Limited access to technology for efficient cropping	Precision Agriculture and farm management
Uneven quality and lack of large scale testing	Quality management and traceability
Inefficient post-harvest supply chain	Supply chain tech and output market linkages
Lack of access to financial solutions	Financial services

Agritech players are transforming the way agriculture is traditionally being done across all stages of the value chain.

- Players in market linkages: farm inputs segment are seamlessly blending technology with physical infrastructure to offer farm inputs at a greater price certainty
- Precision agriculture and farm management players are helping farmer improve their yields by up to 30 per cent
- Quality management and traceability players are helping farmers realize better realizations by incentivizing high quality produce
- Players operating in supply chain tech and output market linkages segment are eliminating inefficiencies such as high wastage of farm produce, which is a win-win for both farmers as well as consumers
- Financial services players could serve 30 per cent of farmer households through access to credit, and 65 per cent of farmer households through access to crop insurance
- Business models in the agritech space could be trifurcated into:
- Margin-based models where players create market linkages on the input or output side, and earn margins on the buy-sell spread
- Subscription-based models where players offer a mix of software, hardware and services to help farmers improve crop yields, track quality of produce or trace the produce across value chain

- ✰ Transaction-based models where players charge based on the number of transactions served such as loans or insurance policies

Demand side drivers such as evolving consumer dynamics towards consuming healthier food due to urbanization, imperative to reduce food wastage in India, environmental factors such as climate change and water shortage are helping drive the adoption of agritech in the country.

Farm automation, often associated with "smart farming", is technology that makes farms more efficient and automates the crop or livestock production cycle. An increasing number of companies are working on robotics innovation to develop drones, autonomous tractors, robotic harvesters, automatic watering, and seeding robots. Although these technologies are fairly new, the industry has seen an increasing number of traditional agriculture companies adopt farm automation into their processes.

Livestock technology can enhance or improve the productivity capacity, welfare, or management of animals and livestock. The concept of the 'connected cow' is a result of more and more dairy herds being fitted with sensors to monitor health and increase productivity. Putting individual wearable sensors on cattle can keep track of daily activity and health-related issues while providing data-driven insights for the entire herd. All this data generated is also being turned into meaningful, actionable insights where producers can look quickly and easily to make appropriate management decisions.

Blockchain's capability of tracking ownership records and tamper-resistance can be used to solve urgent issues such as food fraud, safety recalls, supply chain inefficiency and food traceability in the current food system. Blockchain's unique decentralized structure ensures verified products and practices to create a market for premium products with transparency.

The application of key enabling technologies (KET) is considered the frontiers for innovation in agricultural sector. Farming decisions imply innovation development by the adoption of new and emerging technologies, such as automation and blockchain technology. The process of adopting technological innovations, in fact, is becoming a necessary path, also in agriculture, where it is influenced by a wide range of drivers. Considering that there is a significant increasing of high technology application in agriculture, the technological propensity of agricultural entrepreneurs is becoming necessary to identify and explore new opportunities.

Growth of Agriech Startups in India

The Agritech sector is a fast-growing sector with great potential to benefit Indian agriculture and eventually raise farmers' incomes. Currently, India has 450 Agritech startups and is reported to grow by 25 per cent every year according to NASSCOM, report of 2008.

A wave of Agritech startups has emerged in the last few years in India. In India Agritech sector has come a long way with 43 startups in 2013 to more than 1000 startups in 2020, driven by rising in rural internet penetration, a rise in post-harvest and supply chain losses, a growing number of investors interest in the sector, *etc.* They are now able to solve agricultural problems such as the use of outdated equipment, supply chain management, lack of proper infrastructure, and farmers unable to access a wider range of markets with ease.

Stages of a Startup

There are six stages of a Startup, as given below:

Stage 1- Ideation: The key features are entrepreneurial ambition, initial business models and ideas on how the idea would create value or make money, presence of single person or a vague team, no confirmed roles in the team structure

Stage 2- Concept development: Defining mission and vision of the startup with initial strategy, setting up of key milestones and goals for next few years, Formation of core team, inclusion of some additional members for the requirement specific skill sets along with responsibility

Stage 3- Commitment: Building the MVP (Minimum Viable Product) for the users to test the business idea, Signing of SHA (Shareholder Agreement) among the co-founders

Stage 4- Validation: Most important stage from the point of view of the founder(s), employees, initial customers and angel investors through initial user growth, Initial Key Performance Indicators (KPIs) identified, Founders are struggling to find the right product strategy and brand positioning to attract potential venture investment, Most of the startups lose their plot during this stage of business.

Stage 5- Scaling up: Focus on KPI based growth in terms of users, customers and revenues or market share in the target market, maximum time involved in hiring resources, improvement and distribution of product to target population and implementing new or established process

Stage 6- Establishment of startup: Achieved great growth that can be expected to continue, easy to attract funding and customers at this stage, vision, mission and goals will continue to function "like a startup", Founder(s) or investors may decide to exit or continue with the company.

The investment phases of a startup are given in **Annexure-7**.

6.9.1. AgriTech Startups: Redefining Indian Agriculture through Technology Solution

There is a new wave of budding entrepreneurs and emerging startups in the country that are leading the way to disrupting the agriculture sector in the country. Agritech has the potential to address a number of challenges faced by the sector and, subsequently, change the face of the Indian agriculture.

Fintech platforms connect farmers digitally and provide them financial and agriculture related services based on crops and crop cycle *e.g.* Agritechs such as Samunnati, Jai Kisan belong to this category. Stellapps, Numer8, Aquaconnect, Livestoc, DGV are some of the players who provide platforms to enable lending to livestock farmers.

Farming as a service- Agri startups offer farming services and machinery on rent for reducing capex and increasing affordability *e.g.* Agroapps, CropIn

E-commerce and market linkages- Some startups provide platforms to farmers and traders where they can buy farm inputs and sell farm produce without involvement of middlemen *e.g.* FarmLink, Crofarm

Nanotechnology- There are startups which provide facilities such as nano capsules, nano particles to enhance nutrients uptake and treat diseases *e.g.* JK Nanosolutions, Carbon Continuum

IoT and Big data- Startups facilitating data collection and decision making using drones, sensors and data analytics. *e.g.* Fasal, Intello Labs

Post-harvest technologies- Startups offering technologies and machineries for grading and processing, cold storage to reduce wastage and increase shelf life *e.g.* Arya, Origo, Star Agri

Smart machines- Startups providing remotely operated machines with greater precision using robotics *e.g.* Unnati Agritech, Intello Labs

Precision farming- Startups facilitating application of precise quantity of inputs at the right time for increasing productivity *e.g.* Aibono, Cropin

6.9.2. Leading Agritech Startups in India

1. **Ecozen solutions**- Based in Pune, this Agritech startup founded in 2010 focuses on developing technology-enabled products to strengthen the farm-to-fork value chain of perishables, with a focus on renewable energy and sustainable development.

   Ecofrost, a device that acts as a portable cold room that maintains a low temperature working on solar power and Ecotron, a pump controller for irrigation that also works on solar power are two products which have become popular among the farmers.
2. **FreshVnF** is a Mumbai based Agritech startup founded in 2018 that uses machine learning (ML) to optimize a farm-to-fork supply chain by connecting farmers with hotels, restaurants, and cafes. The company aims to provide fresh farm produce to the end customer within 16 hours of harvesting.
3. **FIB-SOL Life Technologies**- is an agritech startup founded in 2013 by a team of scholars of IIT Madras. The company's technology is focused

on developing low-cost **bio-fertilizers** that help farmers to improve crop yield and soil quality. The company offers products under GEL and DROPS which provide microbial nutrients to enrich the fertility of farmlands.

4. **Agrostar-** is a Pune-based m-commerce startup that sells agricultural inputs directly to farmers, and has established itself as one of the leaders in the agritech space in India after it raised the highest funding for an agritech startup in the country in 2017. AgroStar's tech platform provides a combination of agronomy advice coupled with service and agri input products that enable farmers to significantly improve their productivity and income.

   AgroStar solves the traditional problems faced around duplication, adulteration and product unavailability by making a wide range of great quality products available to the farmer. The company's value propositions to farmers are- Expert advice, Quality, Customer service, Convenience, Fair price and Easy Accessibility.

5. **Ninjacart -** is India's largest fresh produce supply chain company founded in 2015 and based in Bengaluru, which solve one of the toughest problems in the world through technology. The company connects producers of food directly with retailers, restaurants, and service providers using in-house applications that drive end to end operations. Driven by cutting edge technology, Ninjacart source fresh produce from farmers and sell to businesses, across India. The tech-enabled supply chain, with an always-connected logistics network, helps to lower cost, increase speed and eliminate waste.

6. **Stellaps -** It is an end to end dairy technology solutions company, which is an Internet of Things (IoT) startup with a primary focus on data acquisition and machine learning. Stellapps, which was initially incubated at the rural technology business incubator in IIT Madras, monitors the quantity and quality of milk, storage facilities, creates wearable devices for tracking cattle health, and provides infrastructure for housing cattle.

   The SmartMoo Cloud is capable of supporting data arising out of tens of millions of litres of milk through the milk production, procurement and cold chain flow across millions of farmers. Stellapps's SmartMoo IoT platform acquires data via sensors that are embedded in milking systems, animal wearables, milk chilling equipment and milk procurement peripherals.

7. **CropIn:** Founded in 2010, Bengaluru based CropIn has the objective 'To make any farm traceable and maximizing per acre value'. CropIn's flagship product, SmartFarm, enables big companies to map out the supply chain as well as provide real-time updates on sowing,

weather patterns, pests, *etc.*, to companies as well as farmers who have smartphones and internet connectivity. Other products are Smartrisk, Smartsale and AcreSquare.

8. **Agricx Lab-** Established in 2016 with headquarters in Pune, this startup has developed Agricx certification software which provides a quick, accurate, potable and easy to use quality assessment tool. Agricx is the world's first AI-enabled SaaS stack for entities in the business of producing, trading, storing, transporting, processing or financing of agri produce/commodities.
9. **vDrone-** Founded in 2017 with Bengaluru as headquarters, vDrone provides farmers with actionable data through mapping technology. The firm has developed a software which analyses data collected by deploying drone to provide details to farmers about the health and quality of the land and the crop.
10. **BigHaat-** It is the largest agricultural input marketplace platform in India providing wide choice of quality inputs to farmers at their doorsteps. It is a PAN India platform established in 2015 with Bengaluru as headquarters. It connects input manufacturers with the farmers directly and provide stage- wise technical crop advisory services to farmers.
11. **Intello Labs-** Founded in 2016, the company offers computer vision-based solutions that can determine the quality of fruits and vegetables. The Gurgaon-based company simplifies the evaluation of food for farmers, food packers, and retailers by developing an app to test, grade and analyse the visual quality parameters of agricultural commodities.
12. **Trringo13 and EM3 Agriservices**: They are the two pioneers of farm equipment rental service. They can be called the ubers of the Agriculture sector and have successfully replicated the uberisation of renting farm machinery and tractors in India. Using their services, through a mobile application or a phone call, farmers can rent their required farm machinery on a pay per use basis.

AI Startups in Agriculture

Prospera- a cloud based solution that collects all the data available to farmers. It then connects to a device in the field that analyses the data and suggests desired solutions. Prospera works with a variety of sensors and technology such as computer vision.

Farmbot- is a Smart Agriculture Assistor using internet of things. This has taken precision research to the next level by reaching out to the farming community. The controlling unit consists of Raspberry Pi which controls the entire process of the smart assisting robot. Each hub is the combination with various sensors and gadgets and they are interconnected to one focal server by means of remote correspondence modules. The server sends and gets data from client end utilizing web availability.

Harvest CRRO Robotics- harvesting service that automates the crop management, harvesting and packing of specialty crops. The robotics have started with strawberries, that detects and identifies ripe strawberries for harvesting using machine vision and AI.

Gramophone (AgStack Technology)- Gramophone, operated by Agstack Technologies Pvt. Ltd, was launched in 2016 by IIT-Kharagpur and IIM-Ahmedabad alumni. The startup uses image recognition skills to help farmers to get the right information, methods and materials at the right time to get the best harvest possible.

The biggest opportunity for Digital Agriculture across the entire agriculture value chain from the upstream (inputs and production processes) to the downstream (post-harvest and value addition including food processing) is with the emergence of FPOs in India. These FPOs are the vehicle for consolidating farmers, their production as well as the marketing of the produce.

The Ministry of Agriculture in partnership with Startup India launched the Agriculture Grand Challenge in 2017. The challenge was a unique opportunity for Agri-tech start- ups with a commercially viable solution to solve for innovative challenges in the sector.

A Startup which has a profile on the Startup India website is considered a registered Startup on the portal. These Startups can apply for various acceleration, incubator/mentorship programmes and other challenges on the website along with getting an access to resources like Learning and Development Program, Government Schemes, State Polices for Startups, and pro-bono services.

For availing various benefits under the Startup India scheme, an entity would be required to be recognized by DIPP as a startup by applying at https://www.startupindia.gov.in/content/sih/en/startupgov/startup- recognition-page.html.

For DIPP-recognition, the Startups have to apply at

https://www.startupindia.gov.in/content/sih/en/startupgov/startup-recognition-page.html to avail benefits like access to high quality Intellectual Property services and resources, relaxation in public procurement norms, self-certification under labour and environment laws, easy winding of company, access to SIDBI Fund of Funds, tax exemption for three consecutive years, and tax exemption on investment above fair market value.

6.9.3. Institutional Ecosystem of Agri Startups in India

Incubators/Accelerators

A startup incubator is a collaborative program designed to help new startups succeed. The purpose of a startup incubator is to help entrepreneurs grow their business. Startup incubators are usually non-profit organizations, usually run by both public and private entities. Startup accelerators are a type of a business support programme that provide funds, training and mentorship to startups to help them pass successfully through the early stages of business growth. An incubator helps entrepreneurs flesh out business ideas while accelerators expedite growth of existing companies with a minimum viable product (MVP).

An incubator provides array of business services, technology and infrastructure support including office space, mentoring and funding through grants or investor networking opportunities. Accelerators, incubators and mentors identified for the agritech startup ecosystem, along with the pronounced policy and schemes, need to work in tandem with the start-ups to provide the best technical support and reduce their gestation period.

Some of the reputed incubators are:

Madurai Agri Business Incubation Forum (MABIF) is an initiative of NABARD and TNAU, which is an agri-focused incubation centre that only supports Agri startups leveraging technology. It follows regional optimization of local resources for value chain development by aligning the resources to a common point centred on entrepreneurship development.

NAARM has setup an agricultural incubation centre in its campus in Hyderabad captioned **'a-IDEA'** (Association for Innovative Development of Entrepreneurship in Agriculture) with the support of Department of Science and Technology, GOI. Inspired by the success, a-IDEA has now launched relatively more ambitious Phase-II of the programme titled **'Agri Udaan'**. It is India's first Food and Agribusiness accelerator organised by NAARM in association with IIM-A, CIIE in partnership with Caspian Impact Investment and supported by DST. Centre for Innovation Incubation and Entrepreneurship (CIIE) located in IIM, Ahmedabad is a collective of interventions in the space of innovation-driven entrepreneurship in India.

The T-Hub is a unique public-private partnership between the Government of Telangana, three academic institutes-IIIT-H, ISB and National Law University (NALSAR) and key private sector leaders. These entities have come together with a clear vision to create an innovation ecosystem centred around Hyderabad.

Innovation Hub (iHub) was launched by ICRISAT to support agricultural tech entrepreneurs, scientists and technology experts who can collaborate to innovate cutting edge ideas across the whole agriculture value chain.

The MANAGE Incubation Centre under AC-ABC scheme focuses development of startups in the major areas like Agri input, ICT in Agriculture, Animal Husbandry, Nutrition and Health, Farmer Service Centre, Post-Harvest Technology, Farm Mechanization, Supply Chain Management, Fishery and Warehouse Management.

The startups can base their business initiation on incubation centres to handhold them and enable them to run the enterprises for the benefit of the farming community.

MANAGE, Hyderabad in association with Samunnati Financial Intermediation and Services Private Limited, Chennai is in the process of identifying and rewarding successful emerging startups in the country. Samunnati is an Agri value chain solutions provider enabling growth across the agricultural ecosystem, committed to empowering farmers and enabling Agri Enterprises to scale through customised financial and non- financial solutions. It seeks to nurture impact-driven organisations and entrepreneurs in the Agri and agtech space through Samaarambh, startup incubator. IIIT-Hyderabad and MANAGE have signed an MOU to start an Agri tech startup accelerator programme.

Venture acceleration networks consist of experienced, skilled and well-connected individuals who provide hands-on support to entrepreneurs. They help propel viable business ideas to the market place by accelerating the regeneration of ideas and connecting entrepreneurs to the market. They achieve this by: Educating entrepreneurs through on-the-job training on a broad range of practical skills related to the business growth; Connecting entrepreneurs to markets, capital, customers, partners, experts, information and role models through introductions, brokering and by creating bonds of trust and credibility; Validating business ideas through strategic advice and direction, and by creating a supportive environment for business development experiments. Validation provides the critical value-addition of venture acceleration networks. It builds on the two other pillars, educating and connecting, to help ideas fail early, often and inexpensively.

Venture acceleration networks can address a number of market and system failures that hamper innovation and entrepreneurship. These market failures are linked to the transaction costs, search costs and information asymmetries that lead to a less-than-optimal amount of innovation-related services supplied and sold on the market. They are also linked to the positive externalities of investing in the skills of an investee, since part of those skills will be captured by future investors. The system failures are linked to capabilities failures -*i.e.* inadequacies in entrepreneurs' abilities to act in their own best interest - and framework failures related to cultural and social norms.

The public sector can ensure that it quickly converges to appropriate models by supporting existing network initiatives and partnering with regional stakeholders to create new ones. Support can come in the form of facilitating partnerships between relevant stakeholders, providing financing on a competitive basis, and creating mechanisms for programs to adopt good practices and learn from one another.

6.9.4. Government Initiatives and Policies to Promote Startup Ecosystem in India

India has been very active in creating a healthy startup ecosystem, and the growth in the number of startups is increasing year on year. India is among the top five startup communities in the world. The Government of India decided to boost the startup ecosystem in the country and help India become a nation of job creators rather than job seekers. The Government through various initiatives and policies aims to empower startups to grow through innovation and design and to accelerate spreading of the startup movement.

National startup policies have started to show on-ground impact. Multiple enabling policies have been implemented to support startups, their early take off and successful operations (Startup India, 2019).

There are numerous government and semi-governmental initiatives to assist startups.

Some of the policy interventions of the Government of India are:

1. **Start-Up India -** This initiative provides three-year tax and compliance breaks intended for cutting government regulations and red tapism.

   Key international partnerships have been executed by Startup India. Some of them are:

   Indo-Israel Innovation Bridge: It is a tech platform to facilitate bilateral co-operation between Indian and Israeli Startups, tech hubs, corporations and other key innovation ecosystem players. The Innovation Bridge is housed in the Startup India Portal.

   India Singapore Entrepreneurship Bridge is a digital platform to enable Startups, investors and aspiring entrepreneurs of both countries to connect with one another with the focus on: Knowledge exchange, Networking opportunities, and Capacity building

   Sweden-India Startup Sambandh Hub: The digital platform focuses on deepening engagement among stakeholders in both ecosystems. It will provide favourable business environment and necessary guidance to Startups from both the countries for sustainable growth of their businesses.

   Japan India Startup Hub: It is an online platform to bridge the gap between Indian and Japanese startup ecosystems and enable meaningful synergies to promote joint innovation in both economies.

   India Korea Startup Hub: The hub is a one stop platform to bring the Indian and Korean start-up ecosystems closer and to facilitate joint innovation between the two economies. The Hub was conceptualized as part of a joint statement signed between the Korea Trade-Investment Promotion Agency (KOTRA) and Invest India.

The India Tech-Bridge brings together Indian and American entrepreneurs by helping them seamlessly transition to a US presence. The main focus of the program lies in the go-to-market process in the US and providing end to end facilitation for the shortlisted startups. Invest India along with International Accelerator has entered into a strategic alliance to help Indian startups and entrepreneurs succeed by connecting them with the best "in-class incubation support" in India.

2. **Make in India**- The Make in India initiative has been built on layers of a collaborative effort. The Department for Promotion of Industry and Internal Trade (DPIIT) has prepared a road map for the single largest manufacturing initiative undertaken by a nation in recent history. This is helping drive investment, fostering innovation, developing skills, protecting Intellectual Property (IP) and building best-in-class manufacturing infrastructure.
3. **Mudra Yojana**- MUDRA stands for Micro-Units Development and Refinance Agency,which aims to help small scale businesses expand and attain success. Companies from both profit and non-profit sector can avail a loan under this scheme and can avail a loan of up to Rs.10 lakh to kickstart their business. Through this scheme, startups get loans from the banks to set up, grow and stabilize their businesses.
4. **SFAC Venture Capital Assistance**- Venture Capital Assistance is financial support in the form of an interest free loan provided by SFAC to qualifying projects to meet shortfall in the capital requirement for implementation of the project.
5. **E-Biz Portal**- Government launched e-biz portal that integrates 14 regulatory permissions and licenses at one source to enable faster clearances and improve the ease of doing business in India.

6.9.5. Corporate Social Responsibility (CSR) and Agri Startups

India's Companies Act 2013 has introduced the provision for Corporate Social Responsibility (CSR). CSR implies a concept, whereby companies decide voluntarily to contribute to a better society and a cleaner environment. Agricultural development is also included for CSR funding. Every qualifying company requires spending of at least 2 per cent of its average net profit (Profit before taxes) for the immediately preceding three financial years on CSR activities in India. The Agri Startups can utilize CSR funds to scale up its operation and for a wider reach benefitting the farmers.

The evolution of relationship between corporates and society has been one of the slow transformation from hardcore business to a philanthropic and to stakeholder participation. It is observed that only meagre amount of Corporate Social Responsibility (CSR) fund has been spent in the agricultural sector by the companies though there is huge potential for investment in the sector. CSR in the agricultural sector may be viewed in terms of "Creating Shared Value" wherein, business can help progress of agriculture and agriculture sector in turn can help business to improve and flourish.

Some of the CSR Initiatives in agricultural sector are: Zuari Agro Chemicals Ltd, ITC Choupal. ONGC in collaboration with Eklavya foundation has created Centre for Agriculture Research and Training in Telangana for training farmers in organic farming using latest technology. mKRISHI, a mobile application developed by the Tata Trusts with the help of Tata Consultancy Services (TCS) has aided the last-mile technology transfer to the farmers.

REFERENCES

Baljeet Singh (2014) Technology Based Entrepreneurship in Agriculture- Role of Agribusiness Incubator, International Journal of Management and International Business Studies Vol 4, No.3, 2014, p 249-254

Bairwa, S L *et al.* (2014) Agripreneurship Development as a Tool to upliftment of Agriculture, International Journal of Scientific and Research Publications, Vol 4, Issue 3, March 2014

Dana, L. P., and Light, I. (2012). Toward a Theory of Social Capital in Entrepreneurship. *International Journal of Social Sciences, I(1)*, 35–54.

Eric Ries (2011) The Lean Startup, Crown Business, USA

Jonathan, C Ho (2021) Disruptive innovation from the perspective of innovation diffusion theory Technology Analysis and Strategic Management 17 March, 2021. https://doi.org/10.1080/09537325.2021.1901873

Mario Pansera and Richard Owen (2018) Innovation for degrowth: A case study of counter hegemonic practices from Kerala, India, Journal of Cleaner Production, Vol 197, Part 2, October 2018 pp 1872-1883

Pahwa, A (2020) How Agritech startups are changing the face of Indian agriculture. http://www.ey.com>en-in>how-agritech-startups-are.....

Peter Drucker (1985) Innovation and Entrepreneurship, HarperCollins Publishers, India

Rosenkopf, L. and Nerkar, A. (2001) Beyond Local Search: Boundary-Spanning, Exploration, and Impact in the Optical Disk Industry. Strategic Management Journal, 22, 287-306. http://dx.doi.org/10.1002/smj.160

Sarasvathy, Saras (2009) Effectuation: Elements of Entrepreneurial Expertise. Elgar Publishing.

Sharma, CA Vinod Kr (2020) The New Age of Agripreneurship. http://www.ramauniversity.ac.in>news-the-new-age-of-agripreneruship.

Chapter 7

Degrowth Innovations in Agriculture

Degrowth is a movement that aims at achieving a radical change in the societal metabolism of societies, toward a more frugal, sustainable and convivial lifestyle. In a degrowth society, people strive to localise the economies as far and as appropriately as possible, which would assist with reducing carbon-intensive global trade, and building resilience in the face of an uncertain and turbulent future.

Degrowth means primarily the abolition of economic growth as a social objective. This implies a new direction for society, one in which societies will use fewer natural resources and will organize and live differently. Ecological economists define degrowth as an equitable downscaling of production and consumption that will reduce societies' throughput of energy and raw materials.

7.1. Degrowth Movement

The emergence of degrowth as an international research area can be traced back to the first international conference organized in Paris in 2008 by academic collective Research and Degrowth, based at the Autonomous University of Barcelona. At this conference, degrowth was defined as a "voluntary transition towards a just, participatory, and ecologically sustainable society," and proposed as the process that the wealthiest countries should go through in order to achieve a "right-sizing" of both national economies and the global economy. Another five international conferences on degrowth were organized between 2010 and 2016, thus spreading the degrowth movement to several countries around the world.

Degrowth emphasizes the need to reduce global consumption and production and advocates a socially just and ecologically sustainable society with social and environmental well-being replacing GDP as the indicator of prosperity. Well-being, also known as wellness refers to what is intrinsically valuable relative to someone. The well-being of a person is what is ultimately good for this person, what is in the self-interest of this person.

The "degrowth" movement arose from concerns over the perceived consequences of the productivism and consumerism associated with industrial societies, including

- ☆ The reduced availability of energy sources,
- ☆ The declining quality of the environment, including threats to biodiversity and global warming,
- ☆ The rise of negative societal side-effects, including unsustainable development,
- ☆ The exploitative use of resources to satisfy lifestyles that consume more food and energy, and dumbing of greater waste

Scientific evidence and observations increasingly demand a radical change in our current way of life. Degrowth proposes a vision of what such change may mean and places itself in fundamental opposition to the dominant political-economic system. As a movement and academic paradigm, degrowth is a daring and adventurous exploration of alternative ends, values, practices, and structures.

As society does not need another reformation of business which maintains its basic managerial convictions with the concepts of corporate social responsibility, business ethics, sustainable business strategies and social entrepreneurship, degrowth calls for a fundamental transformation of social organisation, economic practices, production and value creation in general.

Degrowth cannot be realised from within a capitalist society, since growth is the sine qua non for capitalism. The ideology of growth is at the core of capitalism. Growthism sustains capitalism politically because it allows avoiding redistribution on the pretext that everyone will continually benefit from it. Growthism pacifies class struggle while justifying existing structures of inequality (Schmelzer, 2016). Today, GDP growth remains the key stabilising mechanism of capitalist economies.

Degrowth is a field of scientific research and activism that starts with the recognition that the flows of material and energy used in the global economy are unequally distributed and often massively oversized, especially in industrialised countries. It advocates that a radical reorganisation and resizing of our economies is needed if equity and sustainability are to be taken seriously. The fundamental degrowth proposition is that it is possible to live well in egalitarian societies with much smaller resource throughputs (Kallis, 2018).

Degrowth theory questions the linchpin of sustainable development: decoupling. By this term environmental economists define the trend by which economic output progressively uses less energy and raw materials because of increases in efficiency. According to this theory, there is no irreconcilable clash between economic growth and environmental preservation.

According to the influential report "Prosperity Without Growth" by economist Tim Jackson (2011), modern economies are built explicitly around consumption growth. The concept was originally released as a report by the Sustainable Development Commission. The report analyses the complex relationships between economic growth, environmental crises and social recession. It proposes a route to a sustainable economy, and argues for a redefinition of "prosperity" in light of the evidence on what really contributes to people's well-being.

7.2. Goals of Degrowth

In academia, the study on degrowth gathered momentum with three main goals:

1. Reduce the environmental impact of human activity;
2. Redistribute income and wealth both within and between countries;
3. Promote the transition from a materialistic to a convivial and participatory society

The first goal is to reduce the environmental impact of human activities through the following proposals:

- ✰ Reduce material and energy consumption;
- ✰ Encourage or create incentives for local production and consumption;
- ✰ Promote changes in consumption patterns.

The second goal is to redistribute wealth both within and between countries through the following proposals:

- ✰ Promote community currencies and alternative credit institutions;
- ✰ Promote a fair distribution of resources through redistributive policies of income;
- ✰ Promote work-sharing and create a citizen's income.

The third goal is to foster the transition from a materialistic to a convivial and participatory society by promoting downshifted lifestyles and by exploring the value of unpaid activity.

7.3. Sources of Degrowth Theories

The following sources are identified as instrumental in development of degrowth theories.

1. **Ecology:** The radical western environmental movement of the 1960s and 70s, particularly with two famous writers, Rachel Carson who wrote Silent Spring in 1962 and Donella Meadows who coordinated 'The Limits to Growth' report in 1972 serve as key figures. This source implies perceiving ecosystems as having value in themselves and also as providers of useful environmental services. Degrowth is considered a possible path to preserve ecosystems by the reduction of human pressure over ecosystems and nature.
2. **Anthropology**: The political, cultural and existential critique of capitalist modernity is evident in the works of Erich Fromm and Ivan Illich. The essence of this source is the critique of uniformization of cultures due to the widespread adoption of particular technologies and consumption and production models expressed in the global North. Erich Fromm points out that 'progress can only occur when changes are made simultaneously in the economic, socio-political and cultural spheres'. Thus, when addressing the microeconomic level, degrowth should never lose sight of the roots, principles, aims and a deep understanding of the interconnection between various societal systems.
3. **Bio-economics:** Ecological economics and industrial ecology are also degrowth sources. Most ecological economists are followers of Georgescu-Roegen (1971), who introduced the term 'bio-economics'. The bioeconomic argument for degrowth including the decreasing energy return on investment and the imminent peak oil are often cited in academic and political debate.
4. **Democracy:** Another source for degrowth movement springs from the need for deeper democracy. Degrowth is a response to the lack of democratic debates on economic development, growth, technological innovation and advancement.
5. **Justice:** Another important source of degrowth is justice. The equality approaches and justice in the context of degrowth implies resource and wealth redistribution both within and between North and South economies.

Degrowth thus goes much beyond the ecological critique of GDP growth. Its answer has to do with concepts like sharing, commons, caring, healing, horizontality, conviviality and simplicity, notions that can all have subversive implications when taken seriously.

Kerschner *et al.* (2018) break down technology for degrowth into classifications of feasibility, viability, conviviality and appropriateness. 'Feasibility' and 'viability' refer to a technology's physical and material structure, exploring implicit and embodied energy consumption of technologies and their sustained impact upon the greater environment.

Alternatively, 'conviviality' and 'appropriateness' extend a social focus and context to degrowth activities. Conviviality draws on 'convivial tools' by Illich (1975), later reinterpreted by Vetter (2018) as 'convivial technologies' to include institutions and products, to recognise the social impacts and relationships emerging from their use. Appropriateness is drawn from Schumacher (1973) where 'appropriate' accords with the values and desires of participants (Alexander and Yacoumis 2018).

Common Misunderstandings about Degrowth

1. First, degrowth is not just about reversing GDP growth rates. This phenomenon is referred to as recession. Degrowth doesn't call for 'less of the same', but for a different way of organising of appropriation, extraction, production, distribution, consumption and waste.
2. Second, degrowth is not about imposing 'austerity' everywhere and shrinking 'everything'. Not everything will be reduced, some of the things will also increase, such as production and consumption of local items, agroecology and urban gardening.
3. Third, degrowth is not anti-technology. However, it is about selecting some technologies to take forward and some to leave behind.
4. Fourth, degrowth is not exclusively local. Small can be beautiful, but not always. Serious post-growth thinking requires a global and systemic approach.
5. Finally, degrowth is not anti-Marxist, but it is against a section of Marxism that has a blind faith in industrial productivism, technological progress, and modern science as the only valid source of knowledge.

7.4. Roots of Degrowth Movement

The key ideas of degrowth are Simone Weil's notion of rootedness within limits, Alexander Chayanov's multiple balances of reproduction, Joseph Kumarappa's democratic decentralisation, Nicholas Georgescu-Roegen's social metabolism, and Joan Martiìnez- Alier's alliance between environmental justice and degrowth.

Simone Weil (1909-1943) was a progressive philosopher and mystic whose revolutionary ideas influenced the twentieth century and beyond. Her philosophy is an early but remarkable example for degrowth. Weil's explanation is an entry point into the relationship between Marxism and degrowth as well as into degrowth's existential critique of industrial modernity, including in the rural world. Her thinking can be seen as a precursor to political ecology and degrowth.

Making one of the earliest critiques of endless growth, she problematised the contradiction between accumulation and the limited availability of natural resources and energy. Her alternative was a vision of 'rootedness' built on autonomous communities and cooperatives, whereby those who work on the land are in charge of their own farms and livelihoods. Weil, who defined uprootedness as the destruction of past and present community ties, argues this severance is necessary for individuals to feel rooted with their surrounding environment in the future.

Weil's notion of 'rootedness' involves the need to be soundly situated in time, space and social relationships. Her idea of rootedness is about 'a re-localisation of effective and affective solidarities'. Rootedness refuses violence and speed, and embraces limits, cooperation, decentralisation, autonomy, friendship and beauty, the material and immaterial 'necessities' of a degrowth society (Azam and Valon, 2016).

Research in degrowth is an attempt to move beyond economistic analyses and to include socio-ecological relations as well as existential questions (Latouche, 2009). Accordingly, degrowth research not only problematises exchange values and abstract labour as potential sites of exploitation, but also critically examines the actual usefulness and implications of 'use values' (as in new technologies and products) as well as the existential nature of concrete labour and community relations.

Alexander Chayanov (1888–1937) was a Russian economist and sociologist who theorised the 'peasant economy'. He was a supporter of peasant households rather than large-scale farms. He argued that the economic logic of the family farm rests on certain balances, particularly between labour and consumption and between utility and drudgery.

Chayanov's analysis of the family farm nourish aspects of degrowth theory and praxis. He noted that because the economy of middle peasants is organised outside the traditional profit/wage concept of capitalism, its economic logic does not purely rely on quantitative necessities. Instead, it crucially depends on qualitative balances defined within the family farm and its context.

Chayanov believed that the peasant economy represents a non-capitalist system that requires its own economic theory. He championed the development of agriculture on the basis of cooperative peasant households instead of the standard soviet policy of large-scale state farms. He also theorised a cooperative-based economic system where different types of production are associated with their appropriate scale. Chayanov's greatest contribution to degrowth is an economy free from the fundamental categories of capitalism and simultaneously based on the popular knowledge of balancing, defining the right scales, and seeking autonomy.

Joseph Cornelius Kumarappa (1892-1960) was an Indian economist, the founder of 'Gandhian economics'. He was in charge of Gandhi's schemes of village reconstruction and helped create and run the All India Village Industries Association (AIVIA), the organisation intended to promote and actualise Gandhi's economic ideas. He promoted a decentralised and self-sufficient economy that would do away with class and caste distinctions while building on the strengths of local production and enabling villagers to control their economic destiny.

Kumarappa considered 'high' vs. 'low' standard of living not as a gradation in the quality of life but a difference between 'complex' vs. 'simple' material standards where 'complex' ones are artificially created by the market and tend to undermine sufficiency and autonomy. Kumarappa was critical of deriving market prices from individual preferences and proposed that valuation standards had to be developed at a deeper level, in the moral economy. Distance between consumer and producers hinders this moral dimension, which is one of the reasons for his defence of local production and consumption.

Nicholas Georgescu-Roegen (1906-1994) was a Romanian-American mathematician and economist considered as the father of ecological economics and a key intellectual influence in the degrowth movement. Georgescu-Roegen suggested a few general principles of agrarian economies. The most relevant to degrowth relates to the agriculturalist's dependence on biotic resources, which have a radically different potential than mineral resources. From this, he developed the central idea of his magnum opus, The Entropy Law and the Economic Process (1971). The common terms of 'production' and 'consumption' obscure the Lavoisian fact that nothing is created or destroyed in the economy, but everything is transformed.

Georgescu-Roegen showed that the mechanisation of agriculture is unsustainably mimicking the industrial logic: machinery, chemical fertilisers and pesticides all rely on mineral resources, rendering modern agriculture, and the industrialised food systems associated with it, as dependent on non-renewable stocks as the industrial sector.

Joan Martiìnez-Alier (1939) is Catalan economist and political ecologist, one of the most influential figures of the degrowth movement. He began his work in Peru to immerse himself in ecological questions, at first through the ecological anthropology of Andean peasants, and later, through the untold history of agricultural energetics and ecological economics in general. Many creative concepts have emerged from his socioenvironmental activism, such as 'ecological debt', 'climate justice', 'biopiracy', 'food sovereignty' and 'land grabbing'.

Martiìnez-Alier emphasises concrete socio-ecological struggles as key sites where 'degrowth' demands are being formulated. For him, such conflicts are 'degrowth in practice', since they de facto seek to prevent growth from taking place, be it in the form of a new highway, a new power plant, a new plantation, or more contamination with industrialisation. Alier has studied agrarian economies, ecological distribution conflicts and indigenous peasants' rejection of modernisation.

In brief, it could be inferred from the degrowth practitioners that Weil stood for localized production, Chayanov suggested basic elements for a degrowth economy, Kumarappa theorised a governance for degrowth, Georgescu- Roegen offered its ecological foundation, while Martiìnez-Alier provided an initial political subject/movement. International Institute of Social Studies (2020) had indicated a deeper analysis of these five older roots of degrowth for setting a new research agenda for agrarian degrowth.

7.5. Degrowth and Technological Innovations

Degrowth and technological innovation have a controversial relationship, some believe they cannot function together, others believe they are supportive. Degrowth has been defined as downscaling production and consumption to improve planetary well-being, ecological conditions, and equity (Schneider *et al.*, 2010). Degrowth calls for a future where societies live within their ecological means; with shrinking economic systems providing greater space for human and ecological ecosystems.

Technological innovation is the act of introducing a new device, method, or material for application to commercial or practical objectives (Schilling and Shankar, 2019). Technological innovations from small-scale community initiatives have the potential to play an important role in urban sustainability and degrowth. The transformative and disruptive role of technological innovation, if developed and applied thoughtfully, can contribute to addressing strong urban sustainability concerns- sustainability that is broad, equitable, and generational.

Urban scholars and practitioners need to find ways where technological transitions into socio-environmental justice are not dictated by the pursuit of economic growth, and where top-down enforced urban governance models are replaced with forms of localized, grassroots, and politically dissenting participants (Bunnell, 2015). These localized-grassroots efforts are where small-scale technological initiatives can be used to support the degrowth philosophy. Examples of these small-scale innovations include urban gardening, urban water supply and housing. Each can contribute to degrowth.

7.6. Innovations in Degrowth-Successful Cases

Ibrahim and Sarkis (2020) had presented three evidence based cases that consider different dimensions and elements of degrowth opportunities from urban Egypt. A brief account of the three cases in presented.

1. Sustainable Building Bricks: The Case of Bastoob, Egypt

Bastoob, a local entrepreneurial firm, instead of conventional cement mortar construction, provides innovative, low-cost, and eco-friendly construction solutions that don't require plastering. The innovative interlocking construction block designed by Bastoob allows for the quick and cost efficient construction of housing units and other buildings. Each brick is constructively designed to lock itself into the other bricks around without the use of cement, thereby significantly lowering the construction carbon footprint. Put together, the bricks are aesthetically pleasing and result in stronger facilities. The building systems has been extensively tested for structural strength and durability, as well as for fire, rain, and sound resistance (Martiz, 2018). The blocks are thermally insulating and therefore result in more energy-efficient constructed facilities comfortable without requiring large mechanical energy to regulate indoor temperature. In addition to faster construction, interlocking bricks can be salvaged without damage when the house is being dismantled in the future.

2. Urban Morphology: The Case of Imbaba Agriculture Social Initiative

Imbaba is one of the most visible informal settlements in the area of Giza that resulted from fragmentation of agriculture land. The case of Imbaba represents an innovation to retrofit Imbaba's west side settlements into agriculture productive units using frugal technological solutions. The technology relies on hydroponic systems and develops a community based strategy to organize the trade of crops. Hydroponics is a technique that grows plants without a layer of soil, that is relatively light in weight, water efficient, and mobile. They are able to offer superior aeration of plant roots compared to soil based systems. The project also has a social enterprise element as it is managed through the collaboration of different localized stakeholders. Imbaba's Community Based Organization (ICBO) acts as the focal point and connects different project entities including rooftop farms technicians, developmental agencies, research institutes, and residents.

3. Lab Equipment Sharing: The Case of Scien-up Egypt

Scien-up is an entrepreneurial platform located in Egypt targeting scientific communities in emerging markets in the Middle East and Africa. The scientific communities are mostly associated with urbanized areas. The company was established to democratize access to essential equipment for researchers and innovators, reduce duplication nationally, and free up scarce resources for other investments. Scien-up's model is driven by the characteristics of "sharing economy". It provides access to scientific services and research collaborations by connecting supply and demand sides through an online platform. This platform lists and moderates equipment sharing to target idle resources that could be exchanged between peers. Scien-up's sharing model for democratized access for scientific equipment aims to balance the under-utilized scientific equipment capacities through peer-to-peer exchange databases that allow scientists to share resources using an online platform.

7.7. Business Model Innovation for Degrowth

The COVID-19 pandemic have forced the companies around the world to rethink their ways of doing business, which warrants thinking about post-crisis business models for sustainability, including the concept of degrowth (shifting the focus away from economic growth to quality of life). Ludeke-Freund and Tobias Froese (2021) have introduced business model innovation, combined with sustainability paradigms such as degrowth, as a promising contribution to an innovation and management toolbox for a post-crisis economy.

A post-crisis economy that is more ecologically sustainable and socially just must build on organisations and networks that create value for and with their stakeholders while also protecting their natural environment. For this, the business models must follow sustainability principles such as circularity, inclusiveness, local and green supply chains, and sufficiency.

Sustainable degrowth is a transformative quest towards a more equitable socio-economic setting, as well as a reasonable level of economic throughput that, taken together, improve the quality of life while respecting planetary boundaries (Kallis *et al.*, 2018).

Research on degrowth highlights that our economy includes local sharing, repairing and self-provision, unpaid family care, and many other non-commodified activities, often ignored. An understanding of economic activities beyond merely monetary and reciprocal transactions is essential in order to imagine, creatively experiment with, and implement new solutions to pressing sustainability challenges.

Sustainable business model innovation improves a company's ability to create, maintain, or regenerate natural, social, and economic capital beyond its organisational boundaries by changing the value proposition for its customers and all other stakeholders and/or the way value is created, delivered, and captured.

The case of community supported agriculture is cited as an illustration.

Community-Supported Agriculture

The concept assumes significance in a socio-economic **context** in which citizens are considered as mere consumers of what companies produce, while intermediaries, such as supermarket chains, are positioned in between the producers and consumers. The **problem** is that this form of economic relation and practice does not convey non-material and non-monetary forms of value creation. It disconnects consumers from the conditions of production. Further, it puts producers in fierce competition with each other and, hence, makes the individual vulnerable to poor local conditions (*e.g.*, a dry season). Here, a **solution** is to share the risk with local citizens who pay a fixed yearly amount, regardless of how successful the production is. Moreover, local citizens engage in

Box 7.1

Degrowth movement does not mean the end of consumerism; it means consuming products that have better longevity, producing less waste and consuming less resources. In essence, degrowth means consuming better and more responsibly. This is taking place as consumers in Western economies are incredibly worried about their impact on the environment and have started changing their consumption patterns accordingly.

Sweden's Flygskam or 'flight shaming' movement, which rejected unnecessary air travel, is a prominent example. It had notable success in changing consumer preferences to less polluting travel alternatives like trains.

Companies need to start adapting now with regard to their products and their production process in order to remain competitive as consumers (the demand) change their behaviour; firms (the supply) will have to adapt to this new form of demand. The companies which do this well will sustain a competitive advantage.

Some companies have taken the lead with standard setting in the time of degrowth. Standard setting is about creating new ways to interact with consumers and make these common across an industry.

Companies could also invest in fixing the products they sell to improve their longevity rather than promoting replaceable products. This strategy is very relevant for electronic goods, for instance. Companies might also share their patents when this supports sustainability — other firms in the same industry would have to adopt this kind of innovation, simply to not be left behind.

A degrowth economy can certainly inspire innovation in companies and it should, if firms want to come out on the top of the competition.

Source: https://economictimes.indiatimes.com/news/et-evoke/degrowth-will-accelerate-as-consumers-change-companies-should-adapt-now-thomas-roulet/articleshow/79883040.cms?utm_source=contentofinterest and utm_medium=text and utm_campaign=cppst

local manufacturing or harvesting, build social connections, experience mutual learning processes, and spend leisure time, for instance, in a green environment. Thus, economic relationships are diversified, monetary exchanges are, to a certain degree, deprioritized, and consumers eventually turn into 'community prosumers.' This reduces producers' workload and frees up time to invest in the quality of organisational processes. This can essentially contribute to making business model design processes more democratic and inclusive.

7.8. Research Agenda on Agrarian Degrowth

The ecology of growth is a fundamental area of future research in 'agrarian degrowth'. Studies on degrowth can provide answers to questions like- In what ways has growthism shaped the countryside? To what extent has it lessened/ increased 'poverty'? What kind of 'wealth' does it create and for whom?

Walker (2008) has indicated that India's pattern of dazzling economic growth has been propelled by a powerful "reinforcing mechanism" through which "inequality drives growth and growth fuels further inequality"'. Such mechanisms must be clarified in order to demystify the current common sense that growth ultimately leads to more welfare.

The extractive and commodity frontiers are another big topic for research in 'agrarian degrowth'. These frontiers epitomise the ever shifting 'limits to growth' and they need more theoretical and empirical investigations. Does agroecology represent a way to stay within planetary boundaries? The kind of agriculture that would match a degrowth society needs to be conceptualised more rigorously.

Infante Amate and Gonzaìlez de Molina (2013) suggested a degrowth strategy for agriculture and food based on four 'Rs': 'Re-territorialization of production, re-localization of markets, re-vegetarianisation of diet, re-seasonalisation of food consumption'. Gomiero (2018) argued that biotech-based agriculture such as genetically modified crops are not suitable for a degrowth society.

Research in agrarian degrowth needs to develop appropriate socio-metabolic approaches and reversely, analyses of the social metabolism must be enriched with insights into its politico-institutional determinants which are often grossly under-incorporated (Gerber and Scheidel, 2018). More research on the socio-political metabolism of rural systems would represent the right road for rigorous assessments of what is possible/desirable to maintain over the long term.

In a world where the vast majority of those working in agriculture are 'small' farmers, what can degrowth learn from pre-, semi-, non- or anti-capitalist smallholders? What is the place of 'small' farmers in a degrowth transition? What kind of politico-institutional context would make them 'degrowth-compatible'? These questions can be answered using Cuban experience.

The Cuban experience with 'degrowth' also relates to the role of crises, which needs more attention. Crises of capitalism must be understood, as Moore (2015) emphasised, in the framework of its total conditions of reproduction, including the contradictory relations of capital and nature. Ecological crises are thus likely to also play a major role in generating alternatives.

7.9. Cuban Agriculture as a Classic Case for Degrowth

Cuba massively expanded industrial agricultural techniques using the concentrated power of the planned economy, but has subsequently revived traditional and developed agroecological methods due the changes in the material conditions of production, following the collapse of the U.S.S.R. The collapse of the Soviet Union, and the concomitant dismantling of CMEA (Council for Mutual Economic Assistance), had disastrous impacts on Cuba, a country which relied heavily on the political and economic support of the socialist bloc.

The survival of Cubans is in large part a result of a complete overhaul of Cuban agriculture, and an ecological movement which has made Cuba the world's largest real life model for socialist degrowth.

By the mid-1990s, a major restructuring of state-owned land had begun to occur. Over half of state-owned arable land was redistributed, mainly into Basic Units of Cooperative Production (UBPCs), and hundreds of thousands of Cubans were given land in usufruct - it remained property of the state, but crucially the means of production were owned by members of the cooperatives.

From the depths of economic and agricultural collapse, the island undertook a major transformation of land and authority, embracing agroecological methods, redistributing state-owned farms, opening cooperatives, and supporting the Cuban people's creative, horizontal-led innovations in food production.

Cubans perceive food products as having distinct spatialities (Bayler, 2017). Those spatialities correlate with different scales and social organizations of production. Import stores, ration stores, state produce markets, private produce markets, and self-provisioning plots provide different types of food. Differences in supply are based on the productive needs of the crop balanced against the capabilities of existing production systems and the resources allocated to specified production as prioritized by centralized planning. Spatiality and capability are linked through the ecological, geographic, and climatic composition of Cuba. This spatial allocation of production responsibilities is promising for resilience and degrowth planning.

Degrowth requires a revolution in the productive relations of society, but they must be disrupted and reformed for a purpose. Land nationalization contributed greatly to the country's ability to adopt degrowth practices. Because the state controlled the land, they were in a position to respond rapidly to the sudden crisis in 1989. The state gave free access to people capable and willing to use land, especially any open urban space, to cultivate whatever scale of food possible. Where possible, the state aided producers with training, seeds, traction animals, and tools.

7.10. Natural Farming as a Degrowth Movement

Natural Farming is a chemical-free traditional farming method. It is considered as agroecology based diversified farming system which integrates crops, trees and livestock with functional biodiversity.

In India, Natural farming is promoted as Bharatiya Prakritik Krishi Paddhati (BPKP) Programme under centrally sponsored scheme, Paramparagat Krishi Vikas Yojana (PKVY). BPKP is aimed at promoting traditional indigenous practices which reduces externally purchased inputs. It is largely based on on-farm biomass recycling with major stress on biomass mulching, use of on-farm cow dung-urine formulations; periodic soil aeration and exclusion of all synthetic chemical inputs. According to HLPE Report, natural farming will reduce dependency on purchased inputs and will help to ease smallholder farmers from credits burden.

The BPKP programme has been adopted in State of Andhra Pradesh, Karnataka, Himachal Pradesh, Gujarat, Uttar Pradesh and Kerala. Several studies have reported the effectiveness of natural farming- BPKP in terms of increase in production, sustainability, saving of water use, improvement in soil health and farmland ecosystem. It is considered as a cost- effective farming practices with scope for raising employment and rural development. (The impact of the programme at the field level needs to be studied further).

Zero Budget Natural Farming (ZBNF)

ZBNF is a method of chemical-free agriculture drawing from traditional Indian practices. Subhash Palekar developed ZBNF in the mid-1990s as an alternative to the Green Revolution's methods driven by chemical fertilizers and pesticides and intensive irrigation.

Instead of commercially produced chemical inputs, the ZBNF promotes the application of jeevamrutha- a mixture of fresh desi cow dung and aged desi cow urine, jaggery, pulse flour, water and soil on farmland. This is a fermented microbial culture that adds nutrients to the soil, and acts as a catalytic agent to promote the activity of microorganisms and earthworms in the soil. A similar mixture, called bijamrita, is used to treat seeds, while concoctions using neem leaves and pulp, tobacco and green chillies are prepared for insect and pest management.

The ZBNF method also promotes soil aeration, minimal watering, intercropping, bunds and topsoil mulching and discourages intensive irrigation and deep ploughing.

NITI Aayog has been among the foremost promoters of Mr. Palekar and the ZBNF method. However, its experts have also warned that multi-location studies are needed to scientifically validate the long-term impact and viability of the model before it can be scaled up and promoted country-wide.

The Indian Council of Agricultural Research is studying the ZBNF methods practised by basmati and wheat farmers in Modipuram (Uttar Pradesh), Ludhiana (Punjab), Pantnagar (Uttarakhand) and Kurukshetra (Haryana),

Box 7.2

The High Level Panel of Experts on Food Security and Nutrition (HLPE) is the UN body for assessing the science related to world food security and nutrition. It is the science-policy interface of the Committee on World Food Security (CFS).

HLPE focus on the concepts of transition and transformation. Many transitions need to occur in particular production systems and across the food value chain to achieve major transformation of whole food systems. Both incremental transitions at small scales and structural changes to institutions and norms at larger scales have to take place in a coordinated and integrated way to achieve the desired transformation of the global food system.

The seven PANTHER principles of Participation, Accountability, Non-discrimination, Transparency, Human dignity, Empowerment and the Rule of law should guide individual and collective actions to address the four dimensions (availability, access, utilization and stability) of FSN at different scales.

Source: http://www.fao.org>home>about-the-HLPE-FSN

evaluating the impact on productivity, economics and soil health including soil organic carbon and soil fertility.

NITI Aayog along with Ministry of Agriculture and Farmers Welfare, Government of India had convened several high level discussions with global experts on Natural farming practices. It is roughly estimated that around 2.5 million farmers in India are already practicing regenerative agriculture. In the next 5 years, it is expected to reach 20 lakh hectares- in any form of organic farming, including natural farming, of which 12 lakh hectares are under BPKP.

Agroecology and other Innovative Approaches (HLPE Report, 2019)

Agroecology is a dynamic concept that has gained prominence in scientific, agricultural and political discourse in recent years. It is increasingly promoted as being able to contribute to transforming food systems by applying ecological principles to agriculture and ensuring a regenerative use of natural resources and ecosystem services while also addressing the need for socially equitable food systems within which people can exercise choice over what they eat and how and where it is produced.

Agroecology is a transdisciplinary science, combining different scientific disciplines to seek solutions to real world problems, working in partnership with multiple stakeholders, considering their local knowledge and cultural values, in a reflective and iterative way that fosters co-learning among researchers and practitioners.

Agroecological practices harness, maintain and enhance biological and ecological processes in agricultural production, in order to reduce the use of purchased inputs that include fossil fuels and agrochemicals and to create more diverse, resilient and productive agroecosystems.

There is no definitive set of practices that could be labelled as agroecological. Agricultural practices can be classified along a spectrum and qualified as more or less agroecological, depending on the extent to which agroecological principles are locally applied.

A concise and consolidated set of 13 agroecological principles of agroecology, *viz.*, recycling; reducing the use of inputs; soil health; animal health and welfare; biodiversity; synergy (managing interactions); economic diversification; co-creation of knowledge (embracing local knowledge and global science); social values and diets; fairness; connectivity; land and natural resource governance; and participation have to be effectively applied.

Agroecology is practised and promoted in various locally adapted forms by many farmers and other food system actors around the world, which needs to be documented and mainstreamed.

Five phases have been identified by Gliessman (2007) in making agroecological transitions towards more sustainable food systems. The first three operate at the agroecosystem level and involve: i) increasing input use efficiency; ii) substituting conventional inputs and practices with agroecological alternatives; and iii) redesigning the agroecosystem on the basis of a new set of ecological processes.

The remaining two steps operate across the whole food system and involve: (iv) re-establishing a more direct connection between producers and consumers; and (v) building a new global food system based on participation, localness, fairness and justice. While the first two steps are incremental, the latter three are more transformative.

Conventional views of innovation in agriculture have often focused on the introduction and spread of adoption of new technologies. Recently greater emphasis has been placed on promoting: i) inclusive and participatory forms of innovation governance; ii) information and knowledge coproduction and sharing among communities and networks; and iii) responsible innovation that steers innovation towards social issues.

There are several innovative approaches to SFSs, which can be clubbed in two main categories: i) sustainable intensification of production systems and related approaches (including climate-smart agriculture, nutrition-sensitive agriculture and sustainable food value chains) that generally involve incremental transitions towards SFSs; and ii) agroecological and related approaches (including organic agriculture, agroforestry and permaculture) that some stakeholders consider to be more transformative. While the former category starts from a premise that, to address future challenges, productivity per unit of land needs to increase in a sustainable manner, which is meant by sustainable intensification, the latter emphasizes reducing inputs and fostering diversity alongside social and political transformation focused on improving ecological and human health and addressing issues of equity and governance.

7.11. Concept of Social Innovation from a Degrowth Perspective

The concept of social innovation is very popular in sustainability and transformation research. Broadly speaking, the concept covers new forms of organization, business models, consumption practices, *etc.* that may lead to social change, towards greater sustainability. This is based on the assumption that a social-ecological transformation cannot be reached by techno-economic innovations alone; instead, new social practices are needed to initiate transformation processes that go beyond the techno-economic innovations and to tap the full potential of the latter in the first place.

Social innovation is very much practice-relevant. The concept contributes to an increased reputation of niches and, thus, to an appreciation of grassroots initiatives within civil society. The concept of social innovation has a 'harmonious' bias. Conflicts and questions of power are underexposed. Such disharmonious processes are primarily not carried out with regard to innovations, but rather with regard to exnovations. The term exnovation refers to the intended elimination (or dismantling) of practices, products, technologies, and infrastructure (Gross and Mautz 2015).

The concept of social innovation favours incremental over transformative approaches. The term incrementalism refers to the notion of resolving complex problems via small steps. Incremental approaches avoid radical changes. Instead, the established structures are supposed to be enhanced and optimized as long as possible (Smerecnik and Renegar, 2010). In contrast, transformative approaches aim at a break with hegemonic structures. Practices and strategies of (collective) actors can be classified as transformative if they aim to change social values, norms, and institutions.

7.12. Institutional Innovation and Degrowth: A Case Study

Pansera and Owen (2016) selected Kerala Sasthra Sahithya Parishath (KSSP), a progressive NGO in Kerala, conceived as a people's science movement undertaking many innovative and progressive interventions for a case study on degrowth and analysed their narrative of science, innovation, development and society and its connections to conceptions of degrowth as an example of counter hegemonic practices in a non-Western setting.

The main theoretical dimensions that emerge from the analysis of the data are classified in two overarching aggregate themes: i) Science and technology for social revolution; ii) The politics of technological development.

Box 7.3

For the first time in several years, farmers can have control over water and electricity supply through solar water pumps (SWP). The new technology may also help improve groundwater management. Pradhan Mantri Kisan Urja Suraksha evam Utthaan Mahabhiyan (PM-KUSUM) project intends to give farmers greater financial security as well as more sustainable water access by generating solar electricity on their farms.

By 2022, the goal is to add 30.8 gigawatts (gW) of solar capacity, including roughly two million off-grid pumps. This initiative would help the Government of India achieve the second sustainable development goal mandated by the United Nations that aims to double agricultural productivity and provide adequate wages by 2030.

Distribution companies can save on capital cost by providing farmer cooperative only one main connection. The technology will enable elimination of agricultural power subsidies. The solar pumps can also provide a steadfast secondary source of income for farmers by selling surplus power to the discoms. Farmers will also be motivated to use less power for agriculture.

Source: http://pmkusum.mnre.gov.in

The various initiatives undertaken by KSSP culminated in the creation of an R&D arm of KSSP called the Integrated Rural Technology Centre (IRTC) in 1987. Over almost two decades of activity, the IRTC has developed a wide range of technological artefacts and processes with the intention to upgrade the productivity of the rural communities in Kerala. These activities have been named People's Technology Initiatives (PTIs).

The KSSP case stresses the fundamental importance of locating technologies within a normative and socio-political degrowth frame. This reinforces the ideas that 'degrowth' is a political and not just a technical project and that technologies are in this respect, political artefacts.

The story of the KSSP provides interesting insights for the degrowth community and for post-growth scholars in general *e.g.* degrowth, steady-state and other non-growth focusing alternatives in two ways.

First, it shows a concrete example of an alternative framing of technology as the outcome of a set of linked, normative principles which include and accommodate those intimately connected to those of degrowth in asking 'what kind of society would enable 'technology for degrowth'. The association of the principle of degrowth with others in a broader constellation seems important. A focus on the principle of social justice in the early years of the KSSP activities delivered a number of alternatives to the public/private sectors to empower the rural population through scientific literacy. These activities created momentum to translate scientific knowledge into practical knowledge linked to the principles of self-sufficiency and autonomy. An emphasis on low-tech and low-scale projects was favoured over scale up and horizontal organization and a locally-based, subsistence economy was favoured over centralised entities and profit. The result is a complex ecosystem of experiments, a process of trial and error that still survives in the shade of the dominant paradigm of Indian growth-oriented industrialisation. Two classic examples are popularization of soap making technology among the households and 'smokeless chula' among the rural households.

Second, the KSSP case shows that alternative technological paradigms based on principles aligned with those of degrowth are not only possible but can and co-exist with the hegemonic paradigm. These can create a big impact as reported by thousands of people who have benefited from their low-tech technologies related to energy and environment.

In the middle of the green revolution and the policy of import substitution led by the central government, the KSSP started promoting appropriate/alternative forms of technologies that better served the needs of the rural population in Kerala. The practical applications of this idea are the PTIs, a sort of 'proto innovation system' applied to the Indian rural world. The creation of the PTIs emerged from the conviction that the mainstream thinking only focuses on how to remove the barriers that affect the interactions between research organizations and the practitioners' world in the limited context of public-private relationships.

During the age of neoliberal expansion in the 1980's to till date, the KSSP has continued to promote technological initiatives focused on people and on the creation of alternative forms of subsistence. Kothari (2014) has named this phenomenon as Radical Ecological Democracy. It is an alternative framework of well- being that rejects extractivism and the paradigm of growth while embracing 'new political governance with decentralised decision-making embedded within larger, ecologically and culturally defined landscapes, a new economics that respects ecological limits and democratises both production and consumption, and a new cultural and knowledge-based society that values diversity, collective synergism, and public innovation'.

7.13. Green Growth as Degrowth

In India, the negative relation of growth with environmental indicators got confirmed during the Covid lockdown with all the environmental indicators showing signs of substantial improvement. This can be witnessed with the increase in green cover, improvements in air quality in cities, attainment of better flow regimes in river systems, *etc.*

Another issue emerges from the perspectives of "green growth" and "degrowth". It has already been accepted in many parts of the developed world that the attempts to nurture current levels of growth are proving to be costly due to degradation and depletion of the natural ecosystem. This has not only created a dent on the ecosystem services, but has been a deterrent for the traditional cultural ethos, that respects all forms of life on Earth.

The contention that the consumption levels of the over-consuming classes in the affluent parts of the world are degrading the natural ecosystem hardly needs a reiteration. Developing and the underdeveloped world perceive development as a mono-directional movement of a single variable, GDP, without considering the broader issues of equity and sustainability.

The notion of "green growth" is now accepted as the solution to prevent degradation of the natural environment as a position that can reconcile between the needs of development and conservation goals. The "green growth" hypothesis is contingent upon the decoupling of the use of natural resources and economic growth. This implies that there has to be a factor that can substitute the incremental change in natural resource use for meeting growth targets.

It is misleading to develop growth-oriented policy around the expectation that decoupling is possible. However, it is also noted that GDP has been shown to be a poor proxy for societal wellbeing, something it was never designed to measure.

Green growth is an oxymoron. Growth is necessarily "brown" in colour, with its concomitant pollution. As such, the proponents of both degrowth and green growth concur on one issue: the need and policy prescriptions for meeting conservation goals, environmental protection, maintenance of ecosystem structures and the development of renewables for reducing pollution.

What India needs at this point in time is a holistic development paradigm that plugs in the ecosystem and distribution concerns in development. Interestingly, the small South Asian country, Bhutan, plugs in these concerns in their Gross National Happiness (GNH) framework. While a large democracy like India may not be in a position to replicate that framework, it needs a different configuration in development thinking than the business-as-usual (BAU) approach.

Green growth is considered as "a level of environmental protection which is not being met by current or 'business-as-usual' patterns of growth. The OECD presents potential impacts of green growth initiatives in cities which include an increase in jobs due to the expansion of the green sector, an increase in a city's attractiveness to firms and human capital, a rise in the production of green commodities and services and a rise in the value of urban land (OECD, 2013). These impacts could foster urban sustainable transition. However, green growth is based on the assumption that "technological change and substitution will allow to absolutely decouple GDP from resource use and carbon emission" (Hickel and Kallis, 2020), and as of now there is no clear evidence for this hypothesis.

Unlike in the case of green growth, degrowth is not a consistent and codified paradigm but rather the conflux of various ideas and political actions. However, it 'represents one of the most far-reaching forms of sustainability transitions' (Khmara and Kronenberg, 2020). Degrowth initiatives usually result from grassroots experiments, hence they develop in a bottom-up way. Schneider (2010) gives an overview about the characteristics of the degrowth transformation: It 'involves a reduction of the capacity to produce and consume in a way that is sustainable, balanced, democratic, convivial, ecological, social, positive, cultural, equitable, innovative, diversified, targeted, local and global and transitory'. Considering these attributes, degrowth initiatives could provide for several criteria for urban sustainable transitions. However, there are also doubt about the transformative capacity of degrowth. For example, there is a danger that many degrowth practices will remain on a small scale with 'little potential to contribute to a regime shift' (Khmara and Kronenberg, 2020) and in that sense may lack transformative leadership.

7.14. Sustainable Transitions in Agriculture

Based on three illustrative cases from different parts of India, Prasad and Dutta (2021) indicate how managing transitions towards sustainability require institutional innovations and new intermediaries that build agency, change relations, and transform structures in food systems. The lessons from three diverse geographies and commodities in India are presented- Urban farming initiatives in Mumbai, Conscious consumer initiatives in semi-urban Gujarat for pesticide-free mangoes, and Resource-poor arid regions of Andhra Pradesh. Through these cases, Prasad and Dutta (2021) indicate that beyond the technological solutions, institutional innovations such as urban community-supported farming models, Participatory Guarantee Schemes, and Farmer Producer Organisations (FPOs) can enable sustainable transitions.

Conceptually, there is a need for a radical shift, beyond incremental improvements and technological fixes, to new kinds of socio-technical systems, often called 'sustainability transitions'.

The alternate systems of practice find it hard to gain legitimacy and widespread acceptance, especially from mainstream institutions that create path-dependence and a technological lock-in to, and dominance of, input-centred socio-technical systems.

The concept of a 'niche', inspired by evolutionary biology, highlights spaces of 'radical innovation' in which a small group of actors supports and nurtures innovative approaches and alternate competencies to change the status quo.

The food and agriculture system refers to all the artefacts and activities related to the production, processing, packaging, and consumption of food, along with their socio-economic and ecological outcomes. It has three core elements: (1) food supply chains; (2) food environments; and (3) consumer behaviour.

Case 1: Encouraging Community-Supported Agriculture through Urban Farming Initiatives

Urban farming, broadly understood as the growing and processing of food-related crops and rearing of livestock within or in the vicinity of urban areas has emerged as a common feature of many civic-ecology initiatives.

In recent years, India is also witnessing noticeable trend of urban community farms, with food safety and ecological integrity being a prime motivation. Several startups have facilitated the nascent interest in growing food by offering a host of services and materials to grow rooftop farms. Examples include iKheti, Edible Routes, Homecrop, and Squarefoot farmers. States like Kerala have seen civic bodies initiate sustainable practices by providing kits and services to grow up to 30 vegetable varieties locally.

'Dream grove' is an urban gardening initiative within an 800 square feet public park in Mumbai. The park's earlier neglected condition prompted a local resident (Marie Paul) to convert it into a more welcoming space for the children and elderly in 2018. She formed a local neighbourhood group to maintain the park on a voluntary basis. Two of the group members (Premila Martis and Diipti Jhangiani) had prior experience growing edible plants as volunteers with a city farming group called Urban Leaves.

Though, initially, the Brihan Mumbai Municipal Corporation (BMC) was apprehensive about the idea, the response of locals and the physical transformation of the place prompted them to help with the upkeep of the park. Now, they also seem invested in the idea of such community-led food gardens.

The group expanded their sphere of activities to connect with a nearby weekend farmers' market to procure organic seeds and salvage the leftovers and waste at the end of the day for composting.

Their increased interest and concern regarding the conditions of the food system led Premila to form a 'Growers to Bandra Homes' group during the lockdown imposed in India from March to June 2020, where she actively sourced and procured harvests directly from the farmers or farmer produce aggregators. The network has been gradually expanding, as other members also began introducing more farmer produce in the group through friends or family. The group members also took on the responsibility of buying in bulk from farmers and dividing the produce at their end, so that the cost of the trips from farms to cities could be minimised.

Active collaborations between farmers' markets and voluntary organisations enabled hyper-local supply chains, and even forming alternate means of economic transactions through barter, volunteering time, resources, *etc.*

The case of Dream Grove, an innovation intermediary from an urban metropolis, highlights the potential of community initiatives in creating a demand for sustainable consumption and collective civic action to support farmer groups.

Case 2: Enabling Conscious Consumerism in Semi-Urban Gujarat

Gujpro, a state-level federation of FPOs, was set up in 2015 to help its 29 FPO members, including 45,000 farmers brought together by an earlier participatory natural resource management network, Sajjata Sangh.

Gujpro's main support to its members included the procurement and processing of groundnut, pulses and oilseeds, and trading in cumin. To diversify its portfolio, it also started providing marketing support to mango growers since 2018. Its 'farmer to consumer co-operative shop', Satvik Grahak Bazaar, was launched in February 2019. In the summer of 2019, Gujpro facilitated the sale of carbide-free Kesar mangoes in Ahmedabad with a well-publicised Kesar Mahotsav.

COVID-19 crisis was effectively tackled by the FPO as indicated below:

1. COVID-induced Disruption in Delivery and Supply: On the supply side, Gujpro staff were unable to visit and coordinate procurement, and the onus shifted to the FPOs to maintain quality and ensure that the mangoes were carbide-free, a key differentiator for Gujpro's customers.
2. Leveraging Social Connections to Build a Local Customer Base: A simple Google form for pre-orders was designed, with an explicit request for advance payments to the FPO, which would minimise the marketing risk for the FPO and reduce the hassle of cash handling at delivery. Quick collective action witnessed the identification of a volunteer and coordinator from each institution, who would manage pre-orders and collections as well as smooth, operations in a decentralised manner, even though the customers database was centrally monitored
3. Conscious Consumerism Post-COVID: Conscious consumerism need not just be a metro fad, and residents of smaller towns are willing to pay slightly more for good and safe products with timely information and peer support.

Case 3: Collective Experimentation to Promote Climate-Resilient Farming in a Drought-Prone District of India

Anantapur, a district in Andhra Pradesh, India, is one of the driest and drought-afflicted areas in the country. Here, the farmers shifted from drought-resistant crops like millets and pulses to a monocrop of groundnut. By the 1990s, the yields and soil fertility decreased significantly. Unable to make both ends meet through agriculture, farmers migrated to nearby cities in search of better livelihood options. The key initiatives involved convincing farmers to return to growing millet varieties, and subsequently building a marketing platform to generate a customer base for the products.

To get a fair price for their produce, farmers decided to collectivise under a legal structure such that they could approach the market directly. Under these considerations, the Dharani Farming and marketing mutually aided cooperative was formed in 2008 (Dharani FaM Coop). This initiative empowered farmers to explore the entire value-chain of their produce.

Through cooperative efforts and investment in training local farmer groups, all the member farmers of Dharani Coop shifted to organic farming, which led to a significant decline in input costs. However, organic produce often involves a costly certification for export markets unaffordable for most small and marginal farmers. Building on the network between farmers, Dharani FaM Coop decided to follow the Participatory Guarantee System (PGS) of certification, where a group of farmers certify their fellow farmer group's produce. PGS certification is cost-efficient and highly reliable, as farmers observe the fields daily and the act of peer feedback promotes mutual aid.

To cater to a diversity of customers, with varying buying capacity and different lifestyles, Dharani developed many food items and value-added products, such as breakfast mixes, and millet snacks. Dharani deals with retailers directly to keep the supply chain simple, enabling the organisation to reduce the burden on the consumers and ensuring that farmers get a larger share of the consumer rupee.

Dharani was also able to subvert the concept of big brand retail by creating local systems of fair-trade practices and consumer linkages, which helped them maintain their autonomy while negotiating with the market. Creating such local, resilient systems of responsible supply chain is important because fluctuating and falling commodity prices have adversely impacted the incomes of primary producers.

In the above three cases, unlike conventional approaches that seek to depend on a technological solution or innovation- a better seed, or a digital platform *etc.*- the innovations have largely been institutional. These include new or changed norms of consumption (Dream Grove), production, (Dharani) or newer markets (Gujpro). The cases also map on to different scales of consumption and production, from a small urban community to larger residential societies and FPOs catering to a large number of consumers.

7.15. Conservation Agriculture as a Degrowth Phenomenon

Conservation agriculture and the resource conservation technologies (RCTs) have proved to be highly useful to enhance resource or input-use efficiency and provide immediate, identifiable and demonstrable economic benefits such as reductions in production costs, savings in water, fuel and labour requirements and timely establishment of crops resulting in improved yields. Gupta and

Pathak (2016) have indicated that yields of wheat in heat and water-stressed environments can be raised significantly by adopting the RCTs, which minimize unfavourable environmental impacts, especially in small and medium-scale farms. Zero tillage can allow farmers to sow wheat sooner after rice harvest, so the crop heads and fills the grain before the onset of pre-monsoon hot weather.

Strategic Programmes Large Initiatives and Coordinated Action Enabler (SPLICE) Division, Department of Science and Technology, Government of India has updated the InfoCrop model for assessing the effect of climate change on crop yields and development of adaptation strategies. The Division also has collated and analysed the existing knowledge on climate change adaptation available in various institutes and universities in the country. This will help in developing strategic knowledge for climate change adaptation and can be implemented by the farmer, as part of the degrowth research in agriculture.

7.16. Climate Change and Degrowth Strategy

The impacts of climate change on economic growth are still an issue for debate. Some consider that climate change could be an opportunity for economic growth, while others argue that it will lead to a slow-growing economy, or even degrowth.

ICAR, Ministry of Agriculture and Farmers Welfare, Government of India launched a flagship network project 'National Innovations in Climate Resilient Agriculture' (NICRA) in 2011. The project aimed at strategic research on adaptation and mitigation, demonstration of technologies on farmers' fields and creating awareness among farmers and other stakeholders to minimize the climatic change impacts on agriculture.

The research on adaptation and mitigation covers crops, livestock, fisheries and natural resource management. The project consists of four components *viz.*, Strategic Research, Technology Demonstration, Capacity Building and Sponsored/Competitive Grants.

In the strategic research, the main thrust areas covered are (i) identifying most vulnerable districts/regions, (ii) evolving crop varieties and management practices for adaptation and mitigation, (iii) assessing climate change impacts on livestock, fisheries and poultry and identifying adaptation strategies. Seven climate resilient varieties and 650 district agricultural contingency plans have been developed besides assessing the risk and vulnerability of Indian agriculture to climate change.

Under technology demonstrations, location specific technologies have been demonstrated in 151 climatically vulnerable districts.

State-of-the-art infrastructure facilities have been established by ICAR in the National Agricultural Research and Education System (NARES) across the country to facilitate the climate change research. Unique infrastructure facilities *viz.*, High Throughput Plant Phenomics, Free Air Temperature

Enrichment facility (FATE), Free Air CO_2 Enrichment facility (FACE), CO_2 Temperature Gradient Chambers (CTGC), Gas chromatography, Atomic absorption spectrophotometers, Environmental growth chamber, UV-VIS spectrophotometer, Thermal imaging system, Psychrometric chambers *etc.*

Box 7.4

As part of Doubling Farmers' Income, NITI Aayog (2019) has come out with the slogan 'Grow Solar, Save water' under Energy-Water-Agriculture Nexus. The perverse energy-water-agriculture nexus into which India has been trapped for more than fifty years is well known. Advancements in technology and falling prices, particularly of solar panels, have opened enormous opportunities of not only shifting the nexus into a virtuous cycle but also transforming rural livelihoods. By focusing only on energy and/or water dimension of the solar energy, India may be missing a huge opportunity of increasing farm income, making agriculture climate resilient, and reducing agro-dependence of rural India.

Experience of states in implementing off-grid solar for irrigation, solar cooperatives (Dhundi and Majkuva in Gujarat), sub-station level solar generation (Maharashtra), feeder-level solarization of irrigation (SKY scheme of Gujarat), though in early stages and often being recalibrated, provide useful lessons to be shared to improve the business models for accelerating use of solar in agriculture. Grid-connected solar irrigation for agriculture and rural electricity supply holds immense potential to achieve the trifecta of saving water, doubling farmer income, and saving electricity. An attractive feed-in-tariff can help set the trade-off in favour of water productivity and conservation.

The World Bank has prepared a report titled "Grow Solar, Save Water, Double Farmer Income - An innovative approach to addressing Water-Energy-Agriculture nexus in Rajasthan"

Kisan Urja Surakhsha evam Utthan Mahabhiyan (KUSUM) scheme has three distinct components, each of which is a self-standing scheme:

(A) Sub-station level solar power plant between 0.5 MW to 2.0 MW to be installed through private sector, power utility, or farmers collective,

(B) Off-grid solar irrigation, and

(C) Grid-connected solar that could convert the existing electric tube-wells to farmer-led solar pumps.

The Government of India has set separate target for each of these schemes.

Source: Workshop on Energy, Water, Agriculture Nexus, Niti Aayog and World Bank, 18th Dec 2019, New Delhi.

have been established at various ICAR institutes to facilitate the climate change research. The construction and operation of psychometric chambers have been undertaken for studying the effect of different environmental conditions *viz.*, temperature, humidity, and air movement on livestock, with special reference to cattle and buffaloes, environmental growth chambers with CO_2 and temperature controls and special calorimetric system to study livestock response to heat stress. Custom hiring centres (CHCs) have been established in 121 NICRA villages to ensure availability of farm implements for timely operations. (Source: PIB, New Delhi dated 6th August, 2021).

Grid-connected solar irrigation is a major positive disruption waiting to happen in India. The opportunity is real; what it would take is to strengthen institutional capacity and mitigate payment risk for surplus electricity generated by solar.

Solar Irrigation Pumps (SIP) offer an innovative and promising solution for irrigation as well as the energy-water agriculture nexus. It provides access to uninterrupted free solar power for irrigation during the day time for farmers. Energy pricing and supply conditions are the most important drivers of agricultural ground water demand in South Asia. Free electricity is to be blamed for the over exploitation of ground water from Punjab down to Tamil Nadu, though its negative impacts are limited by restricted hours, unreliable and often night time supply. Solar irrigation pumps offer reliable free solar power during the daytime which can be more lethal, causing overexploitation of ground water resources. So, the design of a promotional policy for solar pumps should also take account water related issues.

Innovations in agriculture related to climate action showcased by Sustainable Innovation Forum (SIF) is given in **Annexure-8**.

7.17. Studies on Degrowth Innovations in Agriculture

Kallis *et al.* (2018) based on review of research on degrowth summarised their findings as below:

1. There is no empirical evidence of absolute decoupling of throughput from economic growth, sufficient decoupling to stay within environmental limits is logically possible, but physically unlikely.
2. Economic growth has long worked as the key dynamic stabilization mechanism for modern, capitalist societies. The end of economic growth without social transformation would lead to destabilization.
3. Capitalism is incompatible with degrowth
4. Planned degrowth is politically unlikely, given established interests and power relations.

While the adoption framework is popular for measuring agricultural innovation, objective and reliable adoption measurements remain problematic and do not recognize the contextual and dynamic decision-making process. The study by Hermans *et al.* (2020) attempted to move beyond the adoption framework and understand: (a) how agricultural decision-making takes place including the knowledge construction, (b) how agriculture is performed in a context of project intervention and (c) how practice adaptation plays out in the context of interacting knowledge. In this study, a method based on the technographic and participatory approach was used to rethink the concept of 'adoption', understand how agricultural decision-making takes place and how the knowledge is constructed after the introduction of Conservation Agriculture (CA) in two Malawian communities.

The findings of the study confirm that farmer decision-making is dynamic, multidimensional and contextual. The authors introduced four lenses that contribute to better understanding complex innovation dynamics: (a) social dynamics and information transfer, (b) contextual costs and benefits, (c) experience and risk aversion, and (d) practice adaptation. Investments should build on existing knowledge and farming systems including a focus on the dynamic decision process to support the 'scaling up, scaling out and scaling deep' agenda for sustainable agricultural innovations.

The theory of change underpinning the common agricultural innovation diffusion model is based on demonstrating benefits through 'demonstration trials' and training lead farmers to become community advocates. The study has shown that social dimensions, including acceptability and group dynamics, play an important role in the farmer decision-making and efficiency of the diffusion model.

Pratt *et al.* (2021) reviewed the behavioural factors in the adoption of USDA innovations. The authors indicated that USDA innovation include innovations not only in production technology and practices for working lands, but also processes and services, risk management products, safety, and resource management. This diversity of innovation reflects the diversity of areas within the purview of the USDA.

The factors related to adoption of innovation are classified into three categories: economic factors, technical feasibility, and behavioural factors. The study was confined to behavioural factors, which include an array of attitudinal, informational, social, and network aspects.

Some of the behavioural factors that extension agencies consider include - using trusted sources to introduce innovations, adapting programmes to support operators with practical or cultural obstacles to traditional programme engagement, and direct efforts to identify which behavioural factors influence adoption outcomes.

The report has drawn the following conclusions:

- ✰ Including explicit consideration of behavioural factors along with economic and technical feasibility considerations may accelerate innovation diffusion as a driver of the U.S. economy. Behavioural factors, such as risk attitudes, environmental attitudes, and social networks interact with economic and technical feasibility factors in the adoption of any innovation. Individual adoption and widespread diffusion of an innovation are context dependent.

- ☆ Adoption and diffusion of innovations requires planning and deliberate consideration of behavioural factors throughout the innovation life cycle. Three practices relevant for all agencies working in the innovation space that would encourage consideration of behavioural factors throughout the process are (1) identification of stakeholders prior to the beginning of a project or program with as much specificity as possible, (2) detection and reduction of process-related barriers, and (3) post-project assessment.
- ☆ Communication and managing the flow of information, both within and between agencies regarding the innovation process, between agencies and stakeholders, and between stakeholders within their networks, is an essential component to innovation adoption.

REFERENCES

Abrol, D. (2005) Embedding technology in community-based production systems through People's Technology Initiatives: lessons from the Indian experience. Int. J. Technol. Manag. Sustain. Dev. 4, 3- 20.

Akhilesh Gupta and H. Pathak (2016) Climate Change and Agriculture in India, Climate Change Programme, Strategic Programmes Large Initiatives and Coordinated Action Enabler (SPLICE) Division, Department of Science and Technology, Ministry of Science and Technology Government of India, New Delhi

Demaria, Federico., Schneider, Francois., Sekulova, Filka and Martinez-Aller, J (2013) What is Degrowth? From an Activist Slogan to a Social Movement, Environmental Values, 22 (2); 191-215

Escobar, Arturo (2015) "Degrowth, Post development, and Transitions: A Preliminary Conversation." *Sustainability Science* 10 (3): 451–462. doi:10.1007/s11625-015-0297-5.

Exner, Andreas and Christian Lauk (2012) "Social Innovations for Economic Degrowth." *Solutions* 3 (4): 45–49.

Hermans D G Thirze *et al.* (2020) Why we should rethink 'adoption' in agricultural innovation: Empirical insights from Malawi, Land Degradation and Development, May 2020 John Wiley and Sons

HLPE on FSN (2019) Agroecological and other innovative approaches for sustainable agriculture and food systems that enhance food security and nutrition, Extract from the Report (June, 2019)

Ibrahim, E Sherwat and Sarkis, Joseph (2020) Technological innovations and Degrowth opportunities from Urban Egypt: Initiating the Discourse, Frontiers in Sustainable Cities 2,

DOI:10.3389/frsc.2020.492866

Illich, I (1973) Tools of Conviviality, Harper and Row, New York

International Institute of Social Studies (2020) Towards a Research agenda for Agrarian Degrowth, Research Brief, August 2020 Erasmus University, Rotterdam

Jackson, T (2011) Prosperity without Growth, Routledge, London

Julien-Franc'ois Gerber (2020) Degrowth and critical agrarian studies, The Journal of Peasant Studies, 47:2, 235-264, DOI: 10.1080/03066150.2019.1695601

Kallis, Glorgos *et al.* (2018) Research on Degrowth, Annual Review of Environment and Resources, 43 (1)

Kothari, A. (2014) India 2100: towards radical ecological democracy. Futures 56, 62-72. http://dx.doi.org/10.1016/j.futures.2013.10.010.

Kassam, A.; Kassam, L (2021) Paradigms of agriculture. In Rethinking Food and Agriculture; Woodhead Publishing: Sawston, UK, 2021; pp. 181–218

Kinfe Asyehegn, Ana Iglesias, Bernard Triomphe, Philippe Pédelahore, Ludovic Temple (2017) The Role of Systems of Innovation in Adapting to Climate Change: The Case of the Kenyan Coffee and Dairy Sectors, Dans Journal of Innovation Economics and Management 2017/3 , p 127- 149

Latouche, S (2009) Farewell to Growth, Cambridge, UK

Ludeke-Freund, F and Froese, T (2021) Degrowth inspires business model innovation for a sustainable post-COVID economy, LSE Phelan US Centre

Mulgan, Geoff, Simon Tucker, Rushanara Ali, and Ben Sanders (2007) "Social Innovation: What It Is, Why It Matters and How It Can Be Accelerated." *Skoll Centre for Social Entrepreneurship*. URL: eureka.sbs.ox.ac.uk/id/eprint/761

OECD (2013) Putting Green growth at the Heart of Development. http://www.oecd.org>dac>environment-development

Pratt Bryan, Sophia Tanner, and Suzanne Thornsbur (2021) Behavioural Factors in the Adoption and Diffusion of USDA Innovations, Research Education, and Economics, Office of the Chief Scientist U.S. Department of Agriculture August 2021

Pansera, Mario and Owen Richard (2016) Innovation for Degrowth: A case study of counter-hegemonic practices from Kerala, India., Journal of Clean Production (2016) 30:1-12

Prasad, Shambu C and Dutta Deborah (2021) Managing Sustainable Transitions: Institutional Innovations from India. *Sustainability* 2021, 13, 6076. https://doi.org/10.3390/su13116076

Schilling, M. A., and Shankar, R. (2019). *Strategic Management of Technological Innovation*. New York, NY: McGraw-Hill Education.

Smith, A., Fressoli, M., and Thomas, H. (2014). Grassroots innovation movements: challenges and contributions. *J. Clean. Prod.* 63, 114–124. doi: 10.1016/j.jclepro.2012.12.025

Stephen R. Gliessman (2007) Agroecology: The Ecology of Sustainable Food Systems, CRC Press, 2007

Suma, T.R.; Groβmann, K (2017) Exclusions in inclusive programs: State-sponsored sustainable development initiatives amongst the Kurichya in Kerala, India. Agric. Hum. Values 2017, 34, 995–1006.

Schmelzer, M (2016) The Hegemony of Growth: The OECD and the Making of the Economic Growth Paradigm, Cambridge University Press, UK

Vetter, A (2017) The Matrix of Convivial Technology- Assessing technologies for degrowth, J. Clean. Production

Walker, K (2008) Neoliberalism on the ground in rural India: predatory growth, agrarian crisis, internal colonization and the intensification of class struggle, Journal of Peasant Studies, 35 (4): 557-620

Chapter 8

Innovation Management and Innovation Intermediaries in Agriculture

8.1. Meaning of Innovation Management

Innovation management is a combination of the management of innovation processes, and change management. It involves the process of managing an organization's innovation procedure, starting at the initial stage of ideation, to its final stage of successful implementation.

It refers to product, business process, marketing and organizational innovation. Innovation management is the subject of ISO 56000 series standards being developed by ISO TC 279. Innovation management is the cornerstone of successful businesses. Disruptive startups like Tesla and Uber have demonstrated that business legacies or organizational size do not guarantee success. One has to be on edge, take risks, and walk along the novel trends to stay sustainable in the long run.

8.2. Key Elements of Innovation Management

The four key elements of Innovation are Collaboration, Ideation, Implementation and Value Creation.

Collaboration

It is a process in which multiple players contribute towards creating new products with customers and suppliers. Collaboration can occur in all aspects of the business cycle, depending on the context.

Collaboration enhances the implementation of innovative ideas and solutions by creating joint ownership, and spreading the risks to a larger group of actors. The dissemination of innovative practices is propelled by collaboration and knowledge sharing across social and professional networks. It is a working practice whereby individuals work together for a common purpose to achieve business benefit. Collaboration enables individuals to work together to achieve a defined and common business purpose.

Collaboration is an essential aspect of innovation because it enables the spread and novel combination of ideas, spreads risk, propagates skills and builds critical mass. Small and medium enterprises (SMEs) are particularly dependent on connections and collaborators for innovation.

Collaboration has the power to spark innovation in the workplace because everyone brings a unique set of knowledge and skills to the table. Working together and embracing these differences gives birth to new ideas through the blending of unique viewpoints.

Ideation

It is defined as the *process of creating, developing* and *communicating* ideas. Ideas are visual, concrete or abstract base elements of a thought (Ben Jonson, 2005). It is the process of forming ideas from conception to implementation, most often in a business setting. Ideation is expressed via graphical, written, or verbal methods, and arises from past or present knowledge, influences, opinions, experiences, and personal convictions.

Ideation and innovation are two closely related, but distinct, concepts. Ideation springs from its route word, "idea". It involves the mental process of brainstorming concepts and theories. Innovation involves the process of turning those newly generated ideas into a manner of doing something new or doing something that is familiar in a faster, better, cheaper, or more convenient manner.

Design is a plan for arranging ideas in such a way as best to accomplish a particular *purpose.* Design thinking is a *human-centred approach* to innovation that draws from a toolkit to integrate the needs of people, the possibilities of technology, and the requirements for business success. (Tim Brown).

While **Invention** is the *creation* of a product, method, or process that is better than a similar product, method, or process which exists today, **innovation** is *the process* of introducing this invention, product, service, or platform, into the world in such a way that it addresses a need *and* creates value to the company and customers.

Implementation

To successfully implement ideas for innovative products and services, there is need to build a culture of innovation in every layer of the company. To be truly innovative, the company needs to invest in organizational capability, workplace culture, and employee capacity for creative thought. Building a culture of workplace innovation takes time, effort and a commitment to:

- ☆ Collaborate across teams and divisions
- ☆ Encourage the search for accidental innovation
- ☆ Look for innovative solutions in non-obvious areas
- ☆ Encourage proactive, solutions-based thinking
- ☆ Tolerate the occasional failure on the way to successful innovation

Value Creation

Value creation happens when a business or organization uses its work and resources to create something of value that is sold to a customer base. In turn, the business earns a profit for what it has created and the customers have a want or need fulfilled

Value can be defined in many ways, such as incremental improvements to existing products, the creation of entirely new products and services, or reducing cost. Businesses seek to create value because their survival, growth and ability to compete in a rapidly changing market depend on whether they innovate effectively.

Value creation is the bedrock of business. It's what sets an organisation apart from other competitors, secures long-term customers, and brings distinct meaning to the brand and its solution. Without creating a value for a business, unique offering will be seen as just another commodity in the eyes of the target market.

Codesigning Innovations

'Codesign' is a generic term referring to any approach by which researchers contribute explicitly to developing innovations together with other stakeholders (or users) at all or most stages of the innovation process by making use of underlying principles. Core codesign principles include: (1) joint planning, implementation, and decision making related to activities aiming to foster innovation and (2) close coordination among stakeholders at the strategic and operational levels, combining scientific, technical, and local knowledge and other resources.

Codesign aims at achieving better articulation between research supply and user-driven demand for problem-solving. It implies that researchers engage systematically with multiple stakeholders in the iterative, adaptive, flexible, and non-linear process of developing innovations.

Research is not necessary for innovation to take place, but often research is an important part of an innovation process. Innovations usually result from a process of net- working and interactive learning among a heterogeneous set of actors, which may include farmers, input suppliers, traders, processors, researchers, NGOs, and government officials.

8.3. Factors Affecting Successful Innovation

The structure of the guiding standard for Innovation Management System (ISO 56002:2019) covers seven key elements. The success factors, to innovate effectively are structured in seven key areas: context, leadership, planning, support, operations, evaluation, and improvement.

1. **Context**: The organization should track external and internal issues and trends in order to identify opportunities and challenges that can trigger innovation activities *e.g.* user preferences, internal capabilities, *etc.*
2. **Leadership**: Based on the understanding of the context, top management should demonstrate leadership and commitment by establishing an innovation vision, strategy, and policy, including the various roles and responsibilities.
3. **Planning**: Innovation objectives, organizational structures, and innovation portfolios should be established based on the direction set by top management and the identified opportunities and risks.
4. **Support**: The support necessary for innovation activities should be put in place, *e.g.* people with the right competences, financial and other resources, tools and methods, communication and awareness creating activities, as well as approaches for intellectual property management.
5. **Operations**: Innovation initiatives should be established in line with the strategies and objectives. Innovation processes should be configured according to the types of innovations to be achieved: identify opportunities, create and validate concepts, and develop and deploy solutions.
6. **Evaluation**: The performance of the Innovation Management System as a whole should regularly be evaluated to identify strengths and gaps.
7. **Improvement**: Based on the evaluation, the system should be improved by addressing the most critical gaps with regards to the understanding of the context, leadership, planning, support, and operations.

8.4. Innovation Intermediary

Innovation intermediary is a concept in innovation studies to help understand the role of firms, agencies and individuals that facilitate innovation by providing various valuable services. They are regarded as crucial actors that can facilitate the innovation process and support sustainable entrepreneurship.

According to Howells (2006), innovation intermediary is an umbrella term that denotes 'an organization or body that acts as an agent or broker in any aspect of the innovation process between two or more parties'. Intermediaries are defined by what they do or which roles they perform rather than by their characteristics.

The literature on innovation intermediaries has been quite fragmented. Howells (2006) has described it as a dispersed field of study not well grounded theoretically.

Intermediary in Innovation and Technological Development

The role of intermediary in innovation and technological development can be traced back to 'middlemen' in the agricultural sector. The role of intermediary in the innovation process has emerged from a number of different sources and research fields over the last 20 years. These include: (a) literature on technology transfer and diffusion, (b) innovation research on the role and management of such activities and the firms supplying them, (c) the systems of innovation literature, (d) research into service organizations and more specifically Knowledge Intensive Business Services (KIBS) firms.

Diffusion and Technology Transfer

The early interest in role of intermediaries in innovation was in the field of diffusion and technology related to 'change agents', who were observed to have a powerful influence on the speed of diffusion and uptake of new products and services by the farmers.

Watkins and Horley (1986) have taken a more prospective look into what intermediaries might do to help the technology transfer process between large and small firms as part of a policy initiative. They indicated the role that such intermediaries could play as: identifying partners in the first place; helping package the technology to be transferred between the two firms; selecting suppliers to make components for the technology; and providing support in making the deal between the firms concerned. Instead of highlighting the role of intermediaries in the diffusion and technology transfer process, here the analysis is more about intermediaries as organizations and what type of activities they are involved in.

Hargadon and Sutton (1997) in their study focus on how brokers, as agents, facilitate the process of knowledge and technology transfer "across people, organizations and industries." They also stressed that brokering is more than just a linking role, which also helps transform the ideas and knowledge being transferred.

Systems and Networks

The systems of innovation (and technological systems) literature in defining a system of innovation have also recognised the existence of intermediary organizations.

Lynn *et al.* (1996) in their study of 'innovation communities' identify a group of organizations that help to link and transform relations within an innovation network or system. These types of organizations would form what Lynn *et al.* (1996) would term 'superstructure' organizations, which act to provide collective goods to their members and help to facilitate and coordinate the flow of information to 'substructure' firms, who are actually producing the innovations.

Intermediaries as Service Organizations

The role of intermediaries and the process of intermediation has also been explored in the context of service activity and service innovation, particularly in relation to the growth of KIBS. Many KIBS firms have close and continuous interactions with their clients which can involve crucial, but largely hidden, functions in supporting innovative change within their client companies.

8.4.1. Innovation Intermediation as a Process

There are two main functions associated with intermediation, namely the information scanning and gathering function and the communication function, both of which might be associated with the 'front end' of innovation intermediation (Lynn *et al.*, 1996).

There are other studies which take a more involved role for intermediaries, by focusing on specific technologies which intermediaries help transfer between firms and organizations. By specifying technologies rather than information, the studies imply that intermediaries have more complete knowledge about the various technological domains in which they operate. The metaphor frequently used here is 'crosspollination' or 'bridging' between previously unrelated or unconnected groups (Bessant and Rush, 1995). Mantel and Rosegger (1987) also see a wider role for intermediaries as standard setters, or as evaluators of a technology after it has been transferred.

8.4.2. Conceptual Issues in Innovation Intermediation

Firstly, the functions of an innovation intermediary are more numerous and diverse than the various studies have implied. The number and diversity appears to be widening along the innovation value chain as organizations providing such functions:

(a) Move upstream (scoping and intelligence) or downstream (IP protection and commercialisation) in relation to the innovation chain into new roles, (b) diversify into new industries or technologies, or (c) shift into new markets overseas. These shifts are because both intermediaries and their clients discover new needs and requirements for their intermediary roles.

Secondly, the de facto assumption in much of the discussion and analysis of intermediaries is that they operate in a simple triadic 'one-to-one-to-one' basis between, for example, a supplier and its customer in some kind of vertical relationship. However, in distributed innovation systems, intermediaries are increasingly involved in more complex relationships, such as 'many-to-one-to-one', 'one-to-one-to-many', 'many-to-one-to-many', or even 'many -to-many-to-many' collaborations, forming both vertical and horizontal relationships in increasingly distributed innovation networks.

The context of the (multiple) relationships and linkage networks of intermediaries are becoming more important.

Thirdly, it is usually assumed that the supplier initiates and uses the intermediary to help supply customers with a new innovation (*i.e.*, that the implied interaction is 'technology becoming more proactive).

Fourthly, innovation intermediaries are, therefore, not only providing immediate, 'one-off' intermediary services to their clients, but are also seeking to offer longer term, 'relational' innovation capabilities to them as well. These collaborations can last for periods of years, not months. They also provide opportunities for the intermediaries to get to know their clients better as well as to gain more lucrative, value-added contracts.

Fifthly, innovation intermediaries were often not only involved in providing mediated innovation services linking their clients with other organizations, but also supplying services direct to their clients on a one-to-one basis, which involved no other interaction with other organizations. Intermediaries therefore can, and do, provide other functions within an innovation system, such as contract research, testing or training work, which have no third party or brokerage function whatsoever.

Due to lack of conceptual grounding, definitions of intermediaries have not been crystallized, and various concepts are used interchangeably. Innovation intermediary has been described using various terms including broker, boundary spanner, and third party. Innovation intermediaries are also described as 'bridgers' and ' change agents'. They are important as the developers of a new invention or technique are seldom connected to their potential users, or to the firms and organisations that have complementary expertise, knowledge and resources. The same applies to potential users of innovations, so that intermediaries are needed to bring organisations and knowledge together to build supply networks and markets.

Intermediaries play a wide range of roles, facilitating the bringing together of various actors at different parts of innovation processes such as ideation, invention, standards making, managing IPR, commercialisation, creating new market segments *etc.* These intermediaries can specialise in different services. The basic functions include process coordination and matchmaking between innovation seekers and potential solution providers, knowledge and finance broking, testing, standardisation, project valuation and portfolio management *etc.* Each of these activities facilitates the exchange and the building of new knowledge, creates opportunities for experimentation, helps the emergence of standards and common goals, and the formation of partnerships. The intermediaries undertake a range of activities that include: scouting potential collaborators, brokering a transaction, mediating, helping find advice, funding, and supporting collaboration.

8.4.3. Functions of Innovation Intermediaries

As we can see from the definition, innovation intermediaries take several roles such as facilitating the process of bringing together multiple parties at a different stage of the process of innovation like ideation, invention, defining the standards, managing IP, creating new market segments, *etc.* These innovation intermediaries can specialize in different services. The basic function includes coordination and match-making between innovation seekers and potential solution providers.

The functions of the innovation system that are completed by innovation intermediaries take wide range such as:

- ☆ **Knowledge development and diffusion**: creation of new knowledge and facilitation of information and knowledge exchange; eg: technological, scientific and market knowledge; this is built through the R&D sections, learning new applications, *etc.*;
- ☆ **Entrepreneurship**: this is the form of creating a new business;
- ☆ **System infrastructure creation**: development and maintenance of the infrastructure system, physical and non-physical infrastructure;
- ☆ **Resource mobilization**: building and attracting resources (human, financial, complementary) relevant to the innovation system; attracting the top skilled people, providing conditions for potential investors and complementary assets (support services, input goods);
- ☆ **Guidance of search**: attracting external actors to the innovation system, to re-direct their search and investment to the system. Also, shift their focus to specific challenges where there are possibilities for growth opportunities;
- ☆ **Market identification and formation:** identified business opportunities and created demand due to the identification of markets and market niches as well as stimulation of local markets;
- ☆ **Legitimation**: Creating and building understanding, support and legitimacy for the innovation
- ☆ **Facilitation/creation of synergies:** this refers to identification and utilization of the synergies within the system. Collaboration and joint projects including product development, R&D, resource development are just a few of the sea of possibilities that is an outcome of the joint

forces. The innovation intermediaries are the facilitators which monitor and help create efficient collaboration between the parties involved.

Basically, the job of intermediaries is to follow the latest trends on the market and integrate them into the existing business model. The study of Innovation Intermediary Organisation (IIOs) has gained great interest, who are perceived as important agents in connecting knowledge supply and demand in the context of National Systems of Innovation (NSI).

Innovation intermediaries execute a broad range of functions in the context of dynamic innovation processes, where, in addition to creating linkages among diverse actors, they also undertake an "animateur" role where they create system dynamism (Howells, 2006).

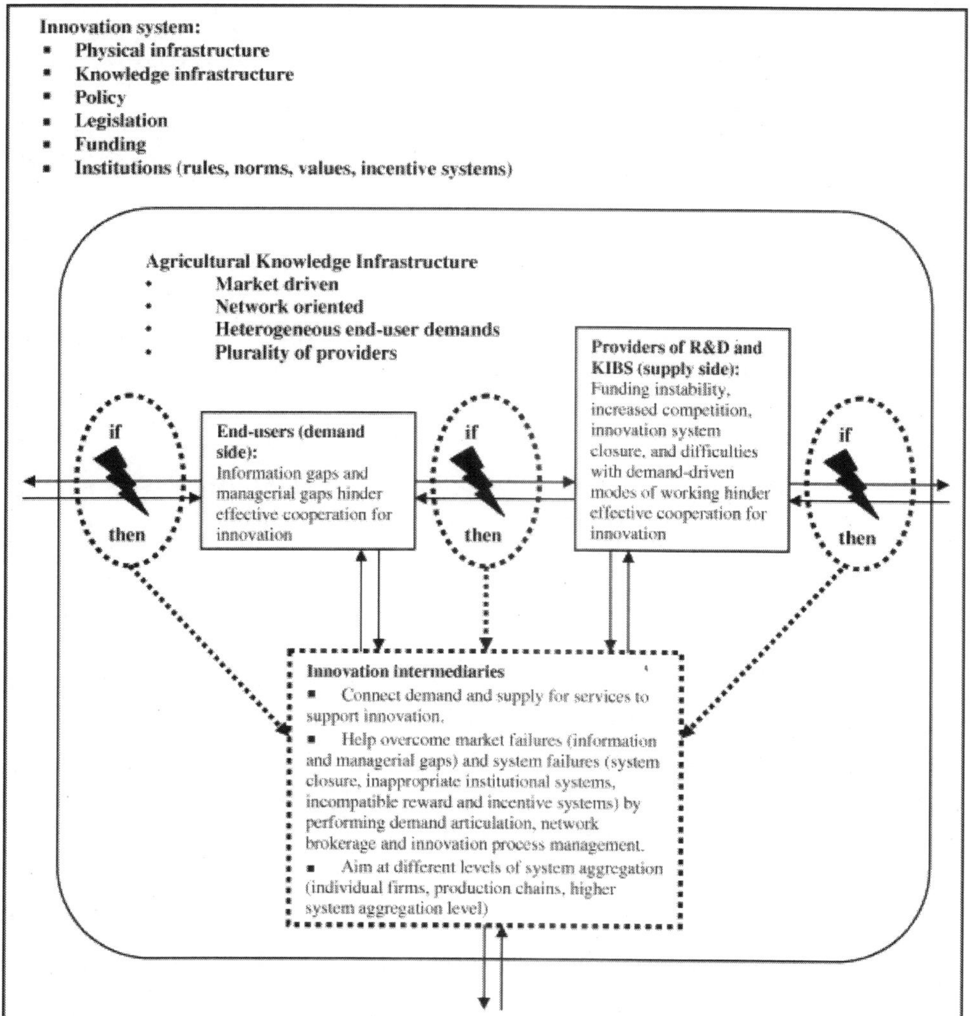

Figure 8: Innovation intermediaries.

8.4.5. Changing Intermediary Domain in Agriculture

Most of the literature on intermediaries in innovation has emerged out of studies in the industrial sector (and increasingly in the health field) that have analysed their role in linking producers and users of scientific knowledge and related technologies in the innovation process. Within this literature, there are different views on intermediaries. One perspective equates intermediaries to knowledge brokers, in the sense of being translators and disseminators of research, much like the classical definition of agricultural extension. However, other scholars distinguish the knowledge broker as one who facilitates access to knowledge, rather than being the expert who is substantively involved in the translation and transmission of this knowledge (Meyer, 2010).

Others have argued that knowledge brokering in principle is not a linear 'science push' process, particularly in increasingly demand-driven approaches to innovation. In the agricultural sector, such knowledge brokering has occurred in the context of emerging knowledge markets in privatized research and extension systems. In this context, the demand side denotes agricultural entrepreneurs, whereas the supply side features R&D and knowledge service providers (Clark, 2002; Klerkx and Leeuwis, 2008). These scholars view knowledge brokering as having the more sophisticated role of matching the demand for and supply of knowledge, entailing articulation of sector innovation visions that then influence research agendas or, at the level of the individual entrepreneur, articulation of demands for farm-specific innovation support services. Further, knowledge brokers have also been understood as intermediaries that occupy 'boundary positions', sitting on the periphery of different worlds and creating an interface between the various actors in innovation networks.

Clearly, in agricultural innovation there is need for knowledge brokering, particularly in a context where sources of knowledge are multiple and highly dispersed (Röling, 2009). However, a sole emphasis on brokering scientific knowledge and technology alone does not take cognizance of the complexity of drivers of agriculture innovation, particularly in developing countries. As Röling (2009) has pointed out, innovation is the emergent property of interaction, and thus the promotion and support of innovation becomes a matter of more broadly facilitating interactions.

In the literature, innovation intermediaries are characterised by a myriad of functions that they undertake in supporting agricultural innovation. Six broad functions include: Demand articulation/stimulation; Network building; Knowledge brokering; Innovation process monitoring; Capacity building and Institutional support.

8.5. Farmers' Organisations as Innovation Intermediaries

The combination of knowledge sources necessitates an active mobilization of diverse actors with multiple perspectives such as farmers, advisory services, agro-companies, *etc.* (Berthet *et al.*, 2016). Mobilizing these actors calls for organizations that act as a bridge to facilitate linkage creation between actors in agroecological innovation.

Examples of innovation intermediaries in the agriculture sector include government agencies, private companies, NGOs, consultants, special programs (*e.g.* research consortia), and Farmers' Organizations (FOs).

Esman and Uphoff (1984) indicated that as intermediary organizations, FOs have now become central in establishing partnerships with public and private agriculture development actors to better access to resources necessary for the provision of knowledge and innovation services. FOs provide various services to members by creating favourable conditions for production and utilization of agriculture knowledge and integrating members into agricultural innovation systems. FOs have been found to implement community-based extension approaches through the creation of joint learning at the individual and organization levels (Kiptot and Franzel, 2019). FOs can also contribute to the creation of innovation platforms so as to allow their members to participate in the co-production of technologies (Pretty *et al.*, 2020).

8.5.1. FPO as an Innovation Intermediary for Better Service Delivery

Collectivization of producers, particularly small and marginal farmers, into farmer producer organisations (FPOs) have emerged as an effective institutional innovation to address their many challenges, most important being improved access to investments, technology, farm inputs and markets.

FPOs enable smallholders to enhance productivity of their crops through efficient and sustainable utilisation of locally available resources and realise higher returns for their produce. The FPOs overcome the problems of unorganised smallholder farmers who lack access to resources and services and have emerged as an interface between smallholders and the external environment by providing forward and backward linkages, giving them a voice, market access, bargaining power and an authority to take timely decisions at their level for their benefit.

In a bid to transform agriculture into a sustainable enterprise, the GOI launched FPOs in the country. The government aims to promote 10,000 FPOs in the next five years, starting 2019-20. The objective of the scheme is to provide small and marginal farmers 'better collective strength for better access to quality inputs, technology, credit and better marketing access through economies of scale for better realisation of income'.

Farmer Producer Organization is a broader term, which includes Farmer Producer Companies (FPCs), farmer cooperatives and farmer societies. Producer Companies (PCs) refer to both farm and non-farm producer companies registered under Producer company provisions of the Companies Act. The term

Box 8.1

National Association for Farmer Producer Organizations (NAFPO) is a stakeholder-driven institutional consolidation of FPO mobilization efforts and works towards creating a support structure for FPOs to identify a pathway towards scale and sustainability. The main objective is to facilitate FPOs to function as collectives, promote the interest of their members while addressing larger challenges such as sustainable agriculture for livelihoods, resource conservation and regeneration, and to promote an equitable and inclusive model of rural economic growth.

Strengthening the FPO Ecosystem: The 4 Pillars of NAFPO Work

1. **Integrating services –** Numerous scattered efforts do exist to address challenges and gaps for promoting FPOs, however, there are very few for stabilizing business and supporting The initiative to integrate services, such as business incubation, statutory compliance support, linkages with financial institutions and buyers on one platform, has been well-received.
2. **Digital ecosystem for FPOs –** With growing interest in the FPO model and the benefits that can be transferred to farmers, a lot of organizations have invested in supporting FPOs. The network of stakeholders – external and internal – is becoming more fluid with transitory structures based on alliances, partnerships, and collaborations. There is a need to develop a digital infrastructure that will benefit FPOs by dynamically aggregating services, exchanging knowledge, finding partnerships, and boosting innovation.
3. **Policy advocacy –** NAFPO organizes multiple surveys, webinars, and consultations to represent the voice of FPOs. The recommendations made during these events are followed up and shared with authorities such as Niti Aayog, NABARD, SFAC, state governments and several ministries. The focus of the recent national and regional discussions was on the AtmaNirbhar Bharat Package, agriculture reforms, and marketing opportunities. NAFPO has also secured a place on the Advisory Committee of both NABARD and SFAC.
4. **Partnerships –** To leverage the network developed by NAFPO, multiple partnerships have been established to facilitate linkages to provide ecosystem support to FPOs. These partnerships cater to challenges faced by FPOs such as access to finance, professional human resource, governance standards, compliance support, technology transfer, customised products like insurance, price risk management and business development support.

Innovations in India, by definition, must conform to four key principles - a new price-performance envelope (so the products and services are affordable to all), scalable (given the size of the country and the market), must be ecologically sustainable, and must represent world-class quality.

Source: https://economictimes.indiatimes.com/view-point/get-to-the-bottom-of-the-pyramid/articleshow/1362084.cms

Farmer Producer Companies (FPCs) refers to producer companies engaged in agriculture and allied activities.

The Producer Companies Act, 2002 was pioneering in its vision of combining the principles of collective action with the structural benefits of a company. It was predicated on the notion that producer companies (PCs) would enable small producers to pool their resources and establish successful businesses that would improve their incomes and reduce risks in the long run.

8.5.2. Value Chain Development by FPOs

At one end of the agricultural value chain are the producers (farmers) and the other end are the consumers. In the middle are several individuals performing one small step in the chain, and each adds value along the way by procuring, processing, transporting, storing, sorting, packaging and retailing.

Under value chains, thousands of scattered small farms are systematically aggregated and provided centralized services around production, post- harvest, and marketing. This helps reduce transaction costs of the farms for accessing the value chains and makes it easier for small farmers to access inputs, technology, and the market. It also opens opportunities to bring primary processing facilities closer to the farm gates and help producers gather market intelligence and manage the value chain better with digital agriculture tools.

The following measures are necessary for the success of Agricultural Value Chain (AVC).

1. Aggregation: helps to address problems related to lack of uniform quality, adoption of sorting and grading technology and resolving appropriate storage needs to infuse efficiency in processing and marketing.
2. Promote good practices and access to timely information to improve bargaining power of farmers vis-aÌ- vis buyers.
3. Address governance issues in agri- food system by giving emphasis on product differentiation, food safety, and product standards that are required in a competitive market environment.
4. Encourage private sector engagement through public-private partnerships for enabling farmers to move up in the value chain.

Agriculture Value Chains as a Strategy for Enhancing Farmers' Income

Value chains act as an important catalyst in improving farmers' income by strengthening the backward and forward linkages of agriculture. This approach has the potential to increase agricultural productivity, household welfare and build social capital. Various innovative strategies are being made to organize farmers and facilitate their access to markets, finances, inputs and technologies.

Case studies in the Indian context show that farmers who participate in value chains incur fewer transaction costs, face lower market risks and realize more profits.

The value chain approach has potential to increase agricultural productivity, household welfare and build social capital (Rutherford *et al.*, 2016). A value chain involves a set of actors and activities that add value to agricultural produce before it reaches to end-consumers. Actors on the value chain include integrators, retailers, lead firms, turnkey suppliers, and component suppliers. It is typically a range of value-adding activities that vertically link interdependent processes to generate value for consumers and establish horizontal links with other value chains that provide intermediate goods and services.

Producers form collectives, *i.e.* associations, cooperatives, producer companies, self-help groups, to derive economies of scale, access new markets and realise higher prices. Although beneficial for producers, they face difficulties in understanding market needs and preferences, and in securing finances from commercial banks and other such institutions.

Consolidators, processors, traders, wholesalers, retailers and exporters develop their own value chains to secure stable supplies from producers. These are termed as buyer driven value chains. Contract farming is the most common form of buyer driven value chain.

The organization of AVCs inclusive of small farmers is an effective strategy to improve efficiency and value addition. AVCs could be seen as an important component of the strategy of doubling farmers' income.

The integrated value chain, that not only connects producers to other actors in the chain (input suppliers, intermediaries, processors, retailers and service providers including warehouse and financial service providers), but also integrates them through ownership and/or formal contracts is also emerging in the farm sector.

Value chain finance is viewed as one of the most sustainable and effective means for financial institutions to improve their outreach to smallholders and other chain actors.

Innovation Intermediaries in the AVC

Agriculture cannot be viewed as farming alone. Instead, it must be viewed as a holistic agriculture value-chain (AVC) system, which includes farming, aggregating, processing, and retailing.

Box 8.2

Bhartiya Agro Industries Foundation (BAIF) facilitated the formation of second-tier marketing organization -Vasundhara Agri-Horti Producer Company Limited (VAPCOL) as a farmers' producer company with 55 institutional members in 2004. Today, this producer organization comprises of farmers from the villages of South Gujarat, Maharashtra, Rajasthan and Madhya Pradesh. VAPCOL acts as a direct link between small, marginal, tribal farmers and the mainstream consumers. Its members are Farmer Organisations (registered or unregistered) having primary membership of individual farmer producers. Farmer Organisations are equity holders in VAPCOL. The main purpose of VAPCOL is of carrying out business procurement, grading, marketing, selling,export of primary produce of the members or procurement of goods and services for their benefit.

The Farm to Kitchen project is initiated by VAPCOL to deliver fresh fruits and vegetables from farmers along with other VAPCOL products to housing societies in Pune and also to tend the demand of Individuals and organizations who would like to donate to the old age homes, orphanages, shelter for homeless labours and community kitchen.

Source: https://vapcol.com

Across this system, there are multiple actors (farmers, traders, retailers, and so on), variables (climate, market prices, and so on), and institutions (seed, fertiliser and other input companies, regulatory bodies, and so on) involved, resulting in dynamic relationships between and within these groups. However, these relationships are often not taken into account when making policies and critical perspectives of farmers are discounted. This leads to increased power imbalances between actors, with marginal and smallholder farmers getting excluded from the AVCs.

AVCs in India, though centred around farmers are influenced by multiple actors at every stage from the production of a commodity on a farm to its final purchase by a consumer. These different actors range from market actors such as wholesalers and agribusinesses to policy- and research-based stakeholders.

All actors in any AVC, along with farmers, need to work together for its smooth operation, since all of them have different expertise, viewpoints and specialised roles that can complement each other, and address the constraints and opportunities that AVCs face.

Disregarding any perspective can lead to the failure of even well-intended interventions. Maya Sen (2021) using two examples from Sikkim and Telangana, explains what can be learned from including multiple perspectives when strengthening any agricultural value-chain.

Sikkim: A Mission Perspective Ignoring Farmers' Conditions

In 2003, Sikkim passed a resolution to completely transit to organic farming. In 2015, it became India's first fully organically farmed state. This achievement was possible due to the inclusive, multi-stakeholder governance in adopting its organic policy. The state focused on involving and aligning stakeholders from the production, technical, policy and marketing activities of the AVCs with the organic mission.

First, Sikkim Organic Mission (SOM) was formed to get the producers (main actors) involved with specific activities. Next, the government regulated the supply of pesticides and chemical fertilisers into the state, pushing for the use of only organic manure. The state government also met all the costs associated with the organic certification process.

Since Sikkimese farmers traditionally grew a large number of their crops organically, shifting to organic farming practices was easier for the local growers. On the marketing side, the government got multiple service providers to focus on actively marketing and spreading awareness about the different organic produce grown in Sikkim. These organisations worked directly with the growers, supporting them improve their agri-management practices and helped organic growers across the state receive better prices for their produce. Further, the organic mission collaborated with banks like NABARD, that supported the creation of Farmer Producer Companies (FPCs) for local growers in East and South Sikkim. Thus, the SOM actively involved multiple stakeholders along the value-chain, so that they could complement one another's expertise and function towards achieving 100 percent organic status.

In the case of large cardamom, the government established auctions in order to make the cardamom market more organised and to help in price protection for farmers. Though large cardamom fetched higher prices in government auctions, the farmers continued to sell to local village-based traders. This is because there was a lack of trust, awareness, and information about the new system. There were also delays in payment and the farmers had limited knowledge of online banking transactions, which was necessary to participate in the auctions.

The auction intervention of the Government failed to become popular because the government assumed certain farmers' needs, and failed to understand and integrate their complete perspective in designing what was a technically well-established system.

Telangana: A Policy Perspective without Proper Planning

Telangana produces paddy in large quantities. In 2020, there was a steep fall in prices of a high-value fine variety of rice (Sannalu). Policymakers in Telangana had recommended farmers to grow Sannalu, as it commands a higher price compared to the regular coarse variety. However, this intervention did not succeed due to faulty planning. The quantity of fine variety grown exceeded the

Box 8.3

Dr Reddys Foundation has developed a community platform called MITRA, which acts as a bridge between farmers and other powerful AVC actors in relaying information. It includes content on best agricultural practices, climate-friendly interventions, farm mechanisation, improved irrigation and water management techniques as well as digital and financial literacy, among other information. Such platforms not only relay information but also take critical feedback from farmers when developing and promoting interventions. This is particularly beneficial, given that agriculture-related information and knowledge is otherwise typically disseminated only through government extension agents, without much room for two-way communication.

Enhanced communication through multi-stakeholder platforms can promote mutual understanding between actors, create a shared vision, help support innovations, and also prevent possible bottlenecks and conflicts that can threaten the smooth functioning of the AVC. Such platforms help actors voice their concerns which can make value-chain collaborations more inclusive and productive.

Source: www.linkedin.com>pulse>mitra-agriculture-inititative

procurement and processing capacity of the rice mills. It was reported that on a single day itself, the capacity of tractors in Miryalaguda (hub of the local rice industry) exceeded 3,000 tractors, as against the actual capacity of 1,500 tractors per day. The mills were not prepared for procuring, storing, and processing such large inflows of fine variety as they were not active participants of the planning process. This led to a situation where the farmers were forced to sell rice at a lower price. Since there were long waiting periods for the trucks carrying the harvest to reach the selling point, large quantities of rice got damaged, further depressing the prices. Such situations often deter farmers from shifting to newer or better crop varieties.

One way to overcome such crisis is by creating multi-stakeholder platforms. These platforms can range from informal governance structures focused on a single sector with a defined geographic scope, to global, formal networks trying to address a wide range of issues.

Roles of Innovation Intermediation in Agriculture

Classical extension has been criticised for its poor capacity to engage multiple interdependent actors (farmers, credit and input providers, buyers, researchers, and development organisations) in joint problem-solving, and in improving farmers' access to broader information relating to market and credit linkages (Klerkx *et al.*, 2012).

They have argued that extension organisations may usefully engage in a broader set of innovation-intermediation roles to enhance the innovation systems performance. Innovation-intermediation relates to three roles described in table below:

Roles of Innovation Intermediaries

Role	*Components*
Demand articulation	Locality-specific stakeholders needs identification Systemic problem diagnosis Collaborative problem assessment
Matching demand and supply	Tailored production advice provision Weather services linkage Transport and tractor services linkage Credit options and linkageInput prices and linkage Market prices and linkage
Innovation process management	Coordination and joint problem solving Maintaining and strengthening relationships Facilitating learning, knowledge integration and co-creation

Source: Adapted from Klerkx et al., 2009.

The roles are:

1. *Demand-articulation*: engaging sector stakeholders to identify their (tacit and expressed) needs (technical, funding or policy), and facilitating the participatory assessment of constraints in meeting their needs and surfacing of interdependencies (Klerkx and Gildemacher, 2012).
2. *Matching demand and supply:* establishing sector contacts (scanning, scoping and filtering actors), identifying potential partnerships that match demand and supply or harness resource complementarity, and developing mutually beneficial relationships (Howells, 2006).
3) *Innovation process management*: creating a discussion and negotiation space for actors to jointly mitigate constraints. This includes maintaining relationships (fostering continuous information flows, trust building and conflict resolution) among stakeholders to cushion mutual benefits (*e.g.*, in credit and markets), and on-going facilitation of knowledge sharing (graduating from learning, knowledge integration to knowledge co-creation) for continuous innovation (Leeuwis, 2010).

Typology of Intermediaries

Systemic Brokers

These intermediaries, who work at higher network level (*e.g.* sector wide), are important in facilitating interactions and coordinating efforts for long-term sector changes. They facilitate demand articulation and options for the desired changes at the system level, and broker networks at the sector level, including industry actors, policymakers, researchers, and government agencies

Specialized Technology Brokers

These brokers work in the realm of emerging agri-biotechnologies and are involved in stimulating demand for new technology and facilitating intricate networks through which knowledge is shared, exchanged, and put into use.

Enterprise Development Support Intermediaries

These intermediaries focus mainly on agribusiness or enterprise development, guided by market demands. Some of these intermediaries work only in the agricultural sector, but a number also have a cross-sectoral focus in supporting small and medium enterprises, including agriculture. These intermediaries focus on facilitating demand articulation for business development services and support network brokering and farmers' capacity building.

Pro-poor Input Access Intermediaries

These intermediaries work in the context of poor households with limited access to knowledge and technologies in predominantly subsistence production systems. This limited access hinders them from improving their production system.

Democratization of Innovation

Innovation is rapidly becoming democratized. Users, aided by improvements in computer and communications technology, increasingly can develop their own new products and services. These innovating users (individuals and firms) often freely share their innovations with others, creating user-innovation communities and a rich intellectual commons. In *Democratizing Innovation*, Eric von Hippel looks closely at this emerging system of user-centred innovation. He explains why and when users find it profitable to develop new products and services for themselves, and why it often pays users to reveal their innovations freely for the use of all. The trend toward democratized innovation can be seen in software and information products, most notably in the free and open-source software movement, also in physical products. Von Hippel's many examples of user innovation in action range from surgical equipment to surfboards to software security features. He shows that product and service development is concentrated among "lead users," who are ahead on marketplace trends and whose innovations are often commercially attractive.

Potential of ICT in Innovation Intermediation

New ICTs have social transformational capacity that positions users as information co-producers rather than consumers and enables online networks. There are high expectations of new ICTs to enhance interaction and information exchange in extension service delivery and other forms of innovation-intermediation. Possibilities include enabling virtual communities for stakeholders to engage in learning and joint problem solving, facilitating virtual connectivity to improve farmers' linkages to services, and the collection of timely and locality-specific information to localise generic scientific knowledge and improve the response to emergent issues

In the sphere of 'demand-articulation', ICTs may facilitate interaction between demand and supply-side stakeholders in various locations on virtual platforms, to engage them in sharing viewpoints, experiences and knowledge relevant to clarifying needs and constraints. They offer opportunities for the creation of timely monitoring systems that support the detection of emerging issues in the field, enabling extension organisations or value chain actors to respond to issues promptly.

This decentralised monitoring of farming activities and data crowdsourcing resonates with forms of citizen science (Cieslik *et al.*, 2017) where information of societal stakeholders may be used to set agendas for innovation and problem-solving. Citizen science is the involvement of ordinary people in scientific knowledge production through engagement in data collection and (or), data analysis (Stevens *et al.*, 2014), therefore facilitating the collection of more observations over a wider geography and expanding possibilities for scientific investigation (Cieslik *et al.*, 2017).

In relation to 'innovation process management', the expectations are that new ICTs (*e.g.*, virtual platforms) may help facilitate continuous interaction, learning and coordination in support of joint problem solving and co-creation of innovation. They may enable the emergence of new forms of organisation in response to environmental challenges or other problems that require collective action.

Bennett and Segerberg have introduced the term 'connective action' to refer to this new form of organisation that is informal and inclusive, and driven by personal motivations to engage with others in the pursuit of change (for instance the Arab Spring). It involves the "articulation of agendas and viewpoints, sharing and shifting of 'memes', information, images and propositions for action, enabled especially by social media such as Twitter and Facebook.

The interest in innovation intermediaries in the agricultural sector is occurring in the context of a shift towards innovation systems approaches, which have directed attention away from the previously dominant linear model that viewed innovation mainly as technology supply. Innovation systems approaches view innovation as a coevolutionary process in which multiple stakeholders interact in dynamic ways to address socio-technical problems caused by many factors in the agricultural sector.

Box 8.4

Unnati is an innovative Fintech company that is revolutionizing the business of short term finance. The company is an online lending startup driven by an app-enabled technology platform to solve the funding challenges of SMEs by offering quick-turnaround, paperless business loans without collateral.

Unnati has joined hands with Farmer Producer Organizations (FPOs) to empower farmers with tech-led business solutions by enhancing their economic strengths and market linkages to improve their income.

Unnati is equipping the FPOs to leverage collective strengths and bargaining power to access financial and non-financial inputs, services, and technologies. This will allow the FPOs to optimise transaction costs and tap into high-value markets. Envisaging the empowerment of small farmers by supporting the FPOs directly via the digital FPO platform, Unnati has plans to do achieve this through digital payment services such as UPI, credit/debit cards, mobile wallets, *etc.*, banking services, output sales, loan services, and brand promotions. The platform will also hold agro advisories and conduct farm predictions to reach its goal of enhancing farmers' lives.

The FPOs, with Unnati's digital solutions, will augment the farmer management, accounting, and inventory management, and will help streamline operations while also bolstering the overall business process. With these FPOs, Unnati aims to ensure farmers have access to a one-stop solution that will provide them with digital solutions to improve their quality of life.

Source: http://unnatiagri.com

Many scholars argue that, in line with the shift towards more demand-driven and pluralistic advisory systems, the role of advisory actors has expanded beyond technology transfer. It now includes organising rural producers, forging links with markets, and brokering multi-actor networks and partnerships in the AIS. The innovation intermediary concept encompasses this vision for a broader and systemic role in supporting innovation processes that is not just about providing extension or advisory services.

Clearly, in agricultural innovation there is need for knowledge brokering, particularly in a context where sources of knowledge are multiple and highly dispersed

Innovation intermediation encompasses broader innovation support and management functions that aim to reinforce relational embeddedness within innovation networks and enhance innovation capabilities. Intermediaries therefore act as 'bridging organisations' that facilitate access to knowledge, skills, services, and goods from a wide range of organisations.

How FPOs can Serve as Innovation Intermediaries in Futures Market?

Farmers, especially small and marginal, do not directly trade in agri-futures market in India. Their small size, lack of trust and understanding of futures market and dependence on middlemen, are some of the main deterrents. The role of FPOs is crucial in this context since they can procure commodities, aggregate them and ensure that size and quality standards required for agri-futures trade are met. Some of the suggestions for better results of FPOs in future markets are:

1. Both FPOs and NCDEX need to focus initially on commodities not perceived by the Government as 'sensitive' from food security point of view so that minimum disruption takes place in futures markets. National Commodity and Derivative Exchange (NCDEX) is the largest agri-commodity exchange in India. This will help them gain confidence in the functioning of futures markets;
2. NCDEX needs to identify production centres, build delivery centres around them and encourage futures trading in these areas;
3. Resource Institutions involved in educating and hand-holding FPOs in futures trading, themselves need to upgrade their knowledge and skills about functioning of futures trading. Government policy and NCDEX both can help them in this direction;
4. Government initiatives like that of Bihar and Rajasthan can help scaling the efforts of linking FPOs to futures markets in other regions,
5. There could also be learning from small holder dominated agriculture of China, that has provided state support in linking farmers to futures, and helped customized products and reduce price distortions,
6. Government's trading arms can also be encouraged to directly participate in the futures market to give confidence to many others, including FPOs; and
7. Instruments like forwards and options have to be encouraged to invite greater participation by FPOs.

Currently, farmers' share in the overall agri-futures trading in India is negligible. There could be several factors behind this ranging from their lack of understanding how futures markets work to constraints emerging from lack of liquidity or even fulfilling regulatory requirements. But linking farmers to futures markets can be mutually beneficial to both- the markets and the farmers. It can help farmers in better price discovery and hedging their price risk while taking planting decisions based on future prices

The first limitation in farmers' participation in futures markets is their small size of the produce itself. It is here that role of FPOs as aggregators become crucial. FPOs can procure and aggregate the produce and ensure that both size and quality standards are met as per requirements for participation in futures markets. This would involve two steps in the process- first, link farmers to FPOs and second, link FPOs to futures markets.

The financial constraint has an impact on the kind of activity the FPOs perform. Most of the FPOs are engaged in business activities of aggregating and selling of inputs like seeds, fertilizer, and pesticides as against aggregation and selling of produce.

While dealing with output related risks, the FPOs have to manage risks related to price volatility, cost of storage and interest costs among many others. Given the financial constraints, FPOs are not equipped and comfortable in handling this risk.

The constraints in linking FPOs to Futures market are- (1) Lack of trust and understanding of futures market, (2) high rejection rates due to mismatch in quality of produce needed by NCDEX and what FPOs provide, (3) reluctance in pre-harvest hedging, (4) logistic related and (5) Tedious documentation and Entry Barriers and (6) stage of growth and maturity of FPOs.

REFERENCES

Bessant, J., Rush, H. (1995) Building bridges for innovation: the role of consultants in technology transfer. Research Policy 24, 97–114

Ben Jonson. 2005. Design ideation: the conceptual sketch in the digital age. Design studies 26, 6: 613-624.

Bennet, W L and Segerberg, A (2012) The logic of connective action: digital India and the personalization of contentious politics, information, Commun. Soc, 15 (2012): 739-768

Catherine Wakesho Kilelu, (2013) Unravelling the role of innovation intermediaries in smallholder agricultural development: Case studies from Kenya, Ph D Thesis, Wageningen University.

Cieslik, K *et al.* (2017) Addressing socio-ecological development challenges in the digital age: environment virtual observatories for connective action, NJAS-Wageningen. J. Life Sci (2017)

Esman, M J and Uphoff, N (1984) Local Organizations: Intermediaries in Rural Development, Cornell University Press, Ithaca, NY.

Hargadon, A., Sutton, R.I. (1997) Technology brokering and innovation in a product development firm. Administrative Science Quarterly 42, 718–749.

Hippel, Eric von (2005) Democratizing Innovation, MIT Press

Howells, Jeremy (2006) Intermediation and the role of intermediaries in innovation, Research Policy, Vol 35, Issue 5, June 2006, p 715-728

Klerkx, L and Gildemacher, P (2012) The role of innovation brokers in agricultural innovation systems, Agric. Innov. Syst. An. Invest. Sourceb. (2012), pp. 221-230

Klerkx, L and Leeuwis, C (2009) Shaping collective functions in privatized agricultural knowledge and information systems: the positioning and embedding of a network broker in the Dutch dairy Sector, J. Agric. Educ. Ext., 15 (2009), pp. 81-105

Klerkx, L, Hall, A and Leeuwis, C (2009) Strengthening agricultural innovation capacity: are innovation brokers the answer? Int. J. Agric. Res. Govern. Ecol., 8 (2009), pp. 409-438

Klerkx, L, Schut, M, Leeuwis, C, Kilelu, C (2012) Advances in knowledge brokering in the agricultural Sector: towards innovation system facilitation, IDS Bull., 43 (2012), pp. 53-60

Lynn, L.H., Reddy, N.M., Aram, J.D.(1996) Linking technology and institutions: the innovation community framework. Research Policy 25, 91–106.

Mantel, S.J., Rosegger, G. (1987) The role of third-parties in the diffusion of innovations: a survey. In: Rothwell, R., Bessant, J. (Eds.), Innovation: Adaptation and Growth. Elsevier, Amsterdam, pp. 123–134

Meyer, AT (2010) Innovate! How Great Companies Get Started in Terrible Times, John Wiley and Sons

Munthali, Nyamwaya., Leeuwis Cees., Paassen Annemarievan, Lie Rico, Asare Richard, Lammeren, Ronvan, Schut Marc (2018) Innovation intermediation in a digital age: Comparing public and private new-ICT platforms for agricultural extension in Ghana, NJAS- Wageningen Journal of Life Sciences, Vol 86-87, 64-76

Singh S (2018) Farm machinery rental services: Case studies from Punjab. In Institutional innovations in the delivery of farm services in India: A smallholder perspective (CMA Publication No. 251). Springer.

Tirtha Chatterjee, Raghav Raghunathan, Ashok Gulati (2019) Linking farmers to futures market in India, Indian Council for Research on International Economic Relations.

Watkins, D and Horley, G (1986) Transferring Technology from large to small firms: the role of intermediaries, In Webb, T *et al.* (eds) Small Business Research, Gower, Aldershot, pp 215-251

Chapter 9

Innovations in Science of Delivery Related to Extension Services

9.1. Introduction

In the earlier chapters, concept of diffusion, disruption and degrowth were covered related to innovations. The present chapter highlights the 'science of delivery' related to innovation transfer and adoption, which is important in the field of agriculture. Science of delivery is the collective and cumulative knowledge base of delivery know-how that helps practitioners make more informed decisions and produce consistent results on the ground. This is very much evident in the use of innovative processes and institutional mechanisms by the extension system.

The seminal World Development Report: 2004 'Making Services Work for Poor People' issued a compelling call for greater accountability to improve the quality of service delivery for the poor. The report indicates that though considerable gains have been made in the intervening decade, despite concerted efforts to promote policy reforms, introduce technical solutions, and allocate additional resources to support these initiatives, many poor people still lack access to basic services, let alone services that meet international standards. There are dire costs from this failure to deliver effective services.

Science of delivery is a disruptive concept that must motivate scientists and development practitioners to collaborate beyond their own disciplines and institutions. Popularizing innovative technologies and achieving larger impacts on the ground requires the involvement of majority of stakeholders. Further, science of delivery require focused funding, institutional incentives, behaviour change, and rethinking on the role of public extension systems (World Bank, 2013 http://blogs.wor1dhank.ors>/developmenttalk/so-what-exactlvis-the-science-of-delivery).

World Bank (2015) in a report titled 'Operationalizing the Science of Delivery Agenda to enhance Development results' lays out a framework for operationalizing the emerging science of delivery. It identifies five delivery principles that practitioners often use to achieve greater impact in their interventions to make more informed decisions.

9.2. Principles of the Science of Delivery

A set of shared characteristics can be identified by analysing successful and unsuccessful projects and what effective delivery efforts have in common. From this, the common characteristics for delivery solutions for citizens could be identified. The five core principles for effective delivery are:

1. Relentless Focus on Citizen Outcomes

Defining the goal as attaining measurable, realistic gains in citizens' well-being; identifying the nature of the delivery problem associated with achieving the gains from both the provider's and the user's perspective; understanding the factors shaping citizens' demands and expectations and the various contextual issues that influence delivery outcomes need to be focused.

In the past, most development projects were designed to focus on generating outputs, rather than the ultimate outcome of producing positive impacts on people (eg: increased yield of crops, not increased income of farmers). As a result, many projects have not consistently achieved the goal of helping the intended beneficiaries. Hence it is necessary to put people and positive outcomes of people at the centre of development projects.

World Development Report (2015) titled 'Mind, Society, and Behaviour' focus on mental and behavioural factors, and sheds considerable light on the issues of problem definition and challenging assumptions, taking into account behavioural approaches to design better solutions. This process applies to the analysis of the preferences and decision-making processes of beneficiaries related to innovations.

2. Multidimensional Response

Facilitating multi-stakeholder coalitions and multi-sectoral perspectives to identify and prioritize problems and possible solutions; convening various development partners and building on their respective comparative advantages need to be adopted as response strategies.

"Hackathons," increasingly becoming popular over recent years, bring together a wide variety of actors to collaboratively develop ideas and form new partnerships. As such, they are a useful tool for achieving multistakeholder coordination and for bringing multiple perspectives and disciplines together to solve problems using innovations. The new connections formed during such events have the potential to bring actors together around innovative approaches, catalysing new initiatives and ways of thinking to achieve impact.

3. Evidence to Achieve Results

The procedure to be followed are: Using the best available evidence to identify problems and solutions; developing local evidence to understand variations in implementation quality (across time, place and social groups) and using it to refine solutions; collecting evidence of results throughout the project cycle to identify weak links; contributing to the global body of knowledge with the evidence collected.

Citizen engagement and participation is an important source of evidence. The Development Impact Evaluation Initiative (DIME) of the World Bank works to build structures of iterative operational learning that put evidence to work by (i) sharing existing evidence at the design stage to introduce potential solutions; (ii) experimentally test project modalities to improve results; and (iii) validating project's overall development hypothesis.

4. Leadership for Change

Understanding local political economies and drivers of change; identifying the incentives that motivate behaviours and integrating these findings into both the design and implementation of delivery solutions; forging durable social compacts between leaders and constituencies to diligently support change through difficult times need to be systematically followed.

The leadership for change principle encompasses three main elements.

1. It is necessary to understand and take into account the political economy, power relations, and political dynamics of the location where we are attempting to effect change. Vital to this process is an understanding of the key players, including beneficiaries and actors who are likely to be obstructive, and the way they are likely to respond to particular initiatives.
2. It is important to understand and align incentives, so that all parties have similar goals, both at the institutional level and often at the individual level.
3. Stakeholders must be convened and brought together within this incentive framework to build and sustain political and developmental coalitions.

5 Adaptive Implementation

There is need to develop an adaptive implementation strategy that allows for iterative experimentation feedback loops and mid-course correction; building a committed team with the right skills, experience, and institutional memory; and maintaining the capacity to reflect on actions and their results.

Crucial for adaptive implementation are small-scale pilot initiatives, with successful efforts subsequently being scaled up. Even successful projects may need to confront failure and change direction to ensure that they ultimately reach a positive impact.

9.3. Innovations in Service Delivery in Agriculture

Innovation arises in a particular socio-economic context and is shaped by the presence or absence of favourable conditions in which it can thrive. The innovation process comes about largely within "innovation systems" made up of organizations and private and public stakeholders interconnected in different ways and possessing the technical, commercial and financial competencies and inputs necessary for innovation.

The concept of innovation systems can be understood in a broad sense and include a wide variety of sectors, including research, extension and other functions that promote or implement innovation. An innovation system consists of a wide array of public and private organizations, firms and individuals that demand and supply knowledge (coded - tacit) and technical, commercial and financial competencies. It also includes the rules and mechanisms by which these different stakeholders interact and relate with one another in social, political, economic and institutional settings (World Bank, 2007).

Box 9.1: Global Delivery Initiative

The Global Delivery Initiative (GDI) of the World Bank was launched in April, 2015. It is a collaboration across the international development community supported by multiple partners to create a collective and cumulative evidence base of delivery know-how to arm practitioners to deliver interventions that result in transformational change. The GDI supports the science of delivery by building on the experience of its partners; connecting perspectives, people, and organizations across sectors and regions; and ensuring that staff and clients have the knowledge they need for effective implementation. The initiative seeks to improve the existing, but fragmented, knowledge of what works in international development- why and most importantly how- and to support practitioners in using these insights to deliver consistent results on the ground.

Source: www.globaldeliveryinitiative.org

Agricultural innovation systems (AIS) are characterized by two main factors: the combination of participants involved, and the dynamic interactions among them. The key participants include farmers and farmer associations, providers of inputs or technical and financial services that promote the development or adaptation of new knowledge, those who encourage an exchange of knowledge and promote learning, those who are engaged in adding value to production, and those who facilitate market access.

Research and technology development organizations are an integral part of the AIS, as are public and private extension services that play a critical role in facilitating access to knowledge and capacity building. A strong science and technology system encompassing basic, strategic, and adaptive agricultural science as well as sciences outside agriculture form the R&D in agriculture, which is widely regarded as contributing to innovation and sustainable, equitable agricultural development.

In the 1950s, innovation in agriculture was essentially thought of as a phenomenon of adoption and adaptation. Science was perceived as external to the socio-economic system, independent and neutral, and a source of innovation, whereas traditional knowledge was seen as a barrier to the spread of progress. In this linear model, support for change consisted of disseminating technological novelties through extension services, which mainly targeted farmers in order to train them in these new technologies.

With more encompassing approaches being required, the discourse among researchers and development agencies gave rise to two new frameworks of thought: AKIS (Agricultural Knowledge and Information Systems) and AIS (Agricultural Innovation Systems) (Klerkx *et al.*, 2012).

The AKIS framework focuses on the exchange of knowledge and information to sustain the innovation process. It is the actors of research and development, education, and agricultural advice who are at the heart of mechanisms for providing support to farmers. The AIS approach is intended to be even more inclusive; it takes into account all the actors who participate, directly or indirectly, in innovation processes (input suppliers, actors of supply chains, banks, policymakers, *etc.*).

Agricultural innovation typically arises through dynamic interaction among the multitude of actors involved in growing, processing, packaging, distributing, and consuming or otherwise using agricultural products. The actors in an AIS innovate not in isolation, but through interacting with other actors: farmers, firms, farmer organizations, researchers, financial institutions, and public organizations and the socioeconomic environment. In other words, agricultural innovation is an organizational phenomenon influenced by individual and collective behaviour, capabilities for innovation, and enabling conditions. Interaction, coordination, and collective action are based above all on the actors' capacity to identify opportunities for innovation, assess the challenges involved, and access the social, human, and capital resources required for innovating, learning, and sharing information.

Like R&D, agricultural extension and advisory services also should be vibrant enough, which have passed through various cycles of challenge and reform. The public services that dominate extension services are plagued by widespread problems: limited funding, insufficient technology, poorly trained staff, weak links to research, and limited farmer participation (World Bank, 2005). Because previous approaches have been ineffective, most extension programmes are moving away from centralized systems and trying to improve links with research and farmers, which have yielded good results globally.

Education and training institutions are especially significant in an AIS because they develop human resources and at the same time serve as a source of knowledge and technology. The absence or decline of these institutions leaves a large gap in a country's innovation capacity. It is in this respect that capacity building institutions are given prominence in promoting innovation.

Agricultural Extension, broadly described as 'the delivery of information inputs to farmers' has traditionally been the primary means of reducing the information asymmetries related to technology adoption in farming in both developed and developing countries (Andersen and Feder, 2007). In general, extension includes transferring technologies from research systems to farmers; advising farm families in their decision-making; educating farmers; and empowering farmers to be able to clarify and realize their goals. To a large extent, most extension programs are publicly funded, supported by local, state, and national governments.

Globally, extension is a key pillar of agricultural development. Though large investments are made in the last few decades to strengthen agricultural research, extension, and advisory services, the extension and outreach systems in most developing countries still remain weak and are often isolated, under-funded, and disconnected from research and education systems.

Extension and advisory services are integral to the AIS, where now more than ever they play a brokering role, linking key actors such as producer organizations, research services, and higher education. Today extension is no longer viewed as an 'agency' but as a 'system' that is integral and central to innovation systems and that focuses on facilitating interaction and learning rather than solely on training farmers.

Although a variety of public and private services are available to farmers, many tasks of extension and advisory services have a public goods nature, including tasks related to regulation, quality control in the produce supply chain, the coordination of service provision, and natural resource management, as well as the provision of services to marginal groups, which are unlikely to access or afford private advisory services.

9.4. Role of Extension Education in Innovations in Service Delivery

Agricultural Extension, also known as Agricultural Advisory Services play a pivotal role in boosting agricultural productivity, enhancing food security, improving rural livelihoods and promoting agriculture as a business enterprise. Extension service provides a critical support system for rural producers meeting the new challenges facing agriculture, transforming the food and agricultural system, including establishment of farmer super markets and establishing standards, labels and food safety measures and promoting agri business by agripreneurs and FPOs.

Modern agricultural extension work covers a wide spectrum of services, which include: improvement in production, marketing, storage, processing, value addition, input supply and distribution, man- power development, irrigation, land management, farm mechanization, *etc.* Thus the scope of present day agricultural extension has expanded well beyond its traditional role in pushing the frontiers of production and productivity. The Doubling of Farmers' Income (DFI) Committee of NITI Aayog, in defining the mandate of Indian Agriculture, describes Agricultural Extension as: "a system of empowering farmers with information, knowledge, technology, skills, risk and farm management practices, across agricultural sub-sectors and along with all aspects of the agricultural supply chain, so as to enable the farmers to realize higher net income from their enterprise on a sustainable basis" (Report XI, 2017).

For doubling farmers' income, the focused areas that demand strengthening of existing extension system identified by NITI Aayog are: delivery of real-time market information; integrating farmers with agri value system platforms; promoting use of warehouses including pledge loan facility and eNWR (negotiable warehouse receipts in electronic form) trade; connecting with agri-logistics, besides using electronic trade platforms; promotion of contract farming; promotion and scaling up FPOs; farm mechanisation; extension of water use efficiency, effective pest management, production and marketing of organic products; primary processing; scientific storage; entrepreneurial skills for farmers to manage small scale agri- business; introduction of new on- farm, off-farm and non-farm income generating activities; intensification and diversification strategies; providing psychological counselling to farmers to manage distresses; developing managerial competencies of farmers and extension personnel and effective use of ICT by farmers.

Several perspectives coincide in the appreciation of learning and knowledge building processes. In today's society, knowledge constitutes a strategic input for development; it promotes creativity and constitutes the raw material for innovation.

Knowledge building depends on the exchange of data, information and ideas. However, knowledge is broader and deeper than data and information, as it is a combination of experience, values, information and knowhow that serve as a framework for the inclusion of new experiences and actions. If knowledge is not exchanged between those who generate, adapt and use it, both wealth and welfare generation becomes more difficult. Hence, not only knowledge from scientific research should be appraised, but also implicit knowledge- both ancestral and recent- from production, distribution and consumption activities for generating significant input for innovation process. The latter knowledge, called tacit, is basically acquired through experience gained along the production process and is represented by the organizational and institutional practices of social and economic agents. It also includes the knowledge possessed by social groups or communities, generated through historical and cultural processes, and who are an integral part of their territorial identity.

Abstract knowledge produced by science and translated into technology and devices, needs supplementary elements: contextual information and associated knowledge. It is dynamic because it regenerates itself and expands through learning.

By the late 1950s, the design of science and technology responded to a linear conception that understood technological development as a unidirectional process (technology push). In the 1960s, 'demand pull' approach emerged within the linear innovation model, modifying the direction of the linear chain.

The determinist linear approach has mainly penetrated the conception of the relation between technology and society, either by considering that technology determines social change –technological determinism- or considering that society determines technology –social determinism. In practice, these theoretical approaches have built a categorical separation between social, environmental, economic and technological problems, constituting different languages that are very difficult to communicate and integrate.

Public extension system in India as well as in many parts of the globe has been criticized as inefficient and ineffective; lacking clear objectives, motivation and incentives; being poorly managed and not accountable to clients; and lacking in relevant technologies (Rivera 1991). Like many other developing countries, the extension services for farmers in India have been publicly funded and delivered through a separate agency, mainly the Department of Agriculture (DoA) of respective states and union territories.

In the 1980s, new innovation process and technological approaches have been developed. The new approaches attempt to simultaneously show the social nature of technology and the technological nature of society, generating a complex and integrative level of analysis, called socio-technical analysis. Some of the characteristics included in the concept of technology, from a socio-technical, systemic and integrative point of view, are:

1. In general, technologies operate within the framework of a historically and territorially defined socio-technical configuration.
2. Understand and assume that all technology is social and that all technology is a cultural expression.
3. Go beyond the static and linear notions of transfer and diffusion to pass on to a transduction process understood as the transformation of a type of energy into another one of a different kind.
4. Go beyond the concept that technologies conceived in developed countries are universal and global. There is no reproducibility but rather re-application. Each solution is useful for each situation, as part of the knowledge implied by it is encoded and another part is tacit.
5. Local dynamics and the strong interactions generated between users and technology re-signify and modify the meaning of technology.
6. Technologies are conceived to actively participate in processes of socio-political, socio-economic and socio-cultural change. They can promote socio-economic development and support democratization processes.
7. The development of technologies with social inclusion implies to develop new modes of complex sociotechnical systems of products, processes and organizations, focused on the dynamics of social and economic inclusion, democratization and sustainable development.

9.5. Models of Extension Service Delivery

Over the past decades, several models of agricultural extension have been developed and implemented to support farmers and rural communities around the world. Every model has its strengths and weaknesses, and the successes of these programmes have varied from country to country.

T and V system of extension was one of the early innovations in agricultural extension. Under the T and V model, systematic training programs were developed and implemented for village extension workers including regular visits to family farms in villages.

The draft "Policy Framework for Agricultural Extension"(PFAE) of the Ministry of Agriculture, GOI emphasised that the "policy environment will promote private and community driven extension to operate competitively, in roles that complement, supplement, work in partnerships and even substitute for public extension" (DAC, 2000). The policy also emphasised to leverage the potential of agri-entrepreneurs, para professionals, custom hiring service providers, input dealers and extension workers in non- governmental organizations to supplement the effort of public extension functionaries. It was felt that capacity building of both public extension functionaries and private extension service providers have to be strengthened for effective service delivery.

The Farmer Field School (FFS) model of Agricultural Extension emerged in Asia in 1980 and was implemented by several institutions and organizations. The FFS model was an intensive, season-long program that focused on experiential learning where farmers met frequently and developed experiments, learned, and shared their skills and knowledge with other farmers in a village.

Krishi Vignan Kendras (KVKs) have emerged as institutional innovations in Agricultural Extension and have become an integral part of the National Agricultural Research System (NARS) in India. KVKs have played an active role in assessment of location-specific technologies for agriculture and allied enterprises, through technology assessment, refinement, and on-farm demonstrations.

The extension approaches pioneered on a small scale by NGOs, FAO, and bilaterally funded projects emphasized participatory learning and action models, with farmer participation and more tailor-made services, including facilitation of access to financial services and access to markets. National and international efforts to revitalize extension brought about a variety of institutional reforms.

In order to address the key constraints faced by extension system in the country with respect to reducing capacity of public extension services, its lack of decentralized and demand driven focus, the Innovations in Technology Dissemination (ITD) component of National Agricultural Technology Project (NATP) was implemented in India. The purpose of NATP's innovation in Technology Dissemination Component is to pilot test new organizational arrangements and operational procedures.

Extension Reforms in India were pilot-tested in 28 districts of seven states during November,1998 to April, 2005 under ITD component of the World Bank funded NATP. This successful experiment served as a basis to launch the Scheme of "Support to State Extension Programmes for Extension Reforms" by the Ministry of Agriculture, GOI to revitalize agricultural extension in all the States to make the extension system decentralized and demand- driven.

One key concept or goal is to decentralize decision-making to the district level through the creation of Agricultural Technology Management Agency (ATMA). The ATMA at district level would be increasingly responsible for all the technology dissemination activities at the district level. It would have linkage with all the line departments, research organizations, non-governmental organizations and agencies associated with agricultural development in the district.

The objectives of ATMA are

1. To strengthen research – extension – farmer linkages
2. To provide an effective mechanism for co-ordination and management of activities of different agencies involved in technology adaption/ validation and dissemination at the district level and below.

3. To increase the quality and type of technologies being disseminated.
4. To move towards shared ownership of the agricultural technology system by key shareholders.
5. To develop new partnerships with the private institutions including NGOs.

With rapid advances in ICTs, e-Extension has become an essential tool in agriculture extension systems and has received enormous attention globally. Access to timely and relevant information is critical to remain competitive in market-driven agriculture. ICTs have created more options for providing advisory services and are increasingly used to circulate market, price, and weather information as well as to offer specific kinds of extension advice.

The Farm Business School (FBS) is an innovative model of agricultural extension services developed by the FAO. This model evolved from the experiences of experiential learning of the FFS model of extension. The FBS model helps farmers to develop their capacities and knowledge in farm businesses, decision-making skills, and entrepreneurship skills.

Agriculture is becoming more technology and information intensive. This transformation is demanding new sets of extension and advisory services requiring new and innovative approaches to serve farmers and stakeholders along the value chains. The recent advances in information and communication technologies are providing innovative tools and approaches for rapidly delivering timely and relevant information to farmers and stakeholders.

The rural democratization process in India has been made possible by the historic legislations on democratic decentralization spelled out in the 73rd and 74th amendments of the constitution that rendered the local self-government institutions mandatory and powerful. While the amendments suggested the broad framework of democratic decentralization and its objectives, the respective state governments were given freedom to improvise the provisions in the amendments and draw up institutional innovations to facilitate the democratization process to reach far and wide. The provisions of the decentralized governance gave considerable freedom to the local bodies to harness the support of institutions and agencies at the local level in innovative ways. For instance, investment in agriculture could be enhanced by seeking partnership from cooperative financial institutions in the locality.

Technology transfer is a critical process in transforming agricultural research innovations into applications for end users. The vast majority of agricultural innovations in developing countries arise from publicly sponsored research centres or universities, which typically are unprepared to engage in formal mechanisms of technology transfer. These mechanisms may require intellectual property protection and/or legal agreements for transferring intangible and tangible property rights to other public, commercial, or international partners.

A sustained investment in capacity building is essential for technology transfer programs to have an impact. It is critical to assess whether an institution has a broad base of research assets and a culture that will support a technology transfer program. If so, investment in a technology transfer office requires paying attention to: (1) identifying an appropriate business model, (2) establishing an institutional policy framework consistent with national laws to clarify responsibilities of the institution and its employees, (3) building the capacity to address intellectual property protection and to negotiate legal agreements to transfer intangible and tangible property, (4) gaining experience to develop business strategies to effectively disseminate technology and work with private partners, and (5) communicating both internally and externally the aspirations (building appropriate culture) and successes of the institution in technology transfer.

9.6. Extension Service as Innovation Brokers

Innovation brokers are persons or organizations that, from a relatively impartial third-party position, purposefully catalyze innovation through bringing together actors and facilitating their interaction. Innovation brokering expands the role of agricultural extension from that of a one-to-one intermediary between research and farmers to that of an intermediary that creates and facilitates many-to-many relationships. As an organization and function,

Box 9.2

Understanding the decisions of farmers to not adopt new varieties and technologies must be better understood. For most humans, "seeing is believing" and hence the role of lead farmers in demonstrating technologies can be critical in their communities. In this regard, Spatial Data Integration (SDI) and India Stack can be leveraged to target the testing of improved technologies with lead farmers. Using real time mobile-based monitoring, the transaction costs and latency of feedback can be dramatically reduced. ICRISAT is now implementing such an approach with the support of an iHub. The ICRISAT innovation Hub (iHub) is a creative space where agricultural technology entrepreneurs, scientists and experts can collaborate to innovate cutting edge ideas for smallholder farmers.

Science of Delivery is focused on how to realize large and sustained impact that draws on knowledge management and diverse methods of sharing that include large-scale demonstrations to inform decisions, adapt approaches and changed mindsets that accelerate the innovation cycle and ensure local conditions, context and culture are considered in developing and delivering products and services. It also helps compress timelines from discovery to delivery.

Source: ICRISAT Happenings Newsletter.

innovation brokering differs from traditional extension and R&D because it represents the institutionalization of the facilitation role, with a broad systemic, multi-actor, innovation systems perspective.

Innovation brokers act as "systemic intermediaries" in innovation systems, forging many-to-many relationships. While the term "broker" has the connotation of a strategically acting go-between who benefits from the separation between actors and pursues objectives mainly out of self- interest, the concept of "innovation broker" derives from the notion of an "honest broker," who brings people together mainly for altruistic purposes. The role of the honest broker resembles a broadened notion of the role of a process facilitator (Klerkx and Leeuwis, (2009). In other words, innovation brokers are facilitators of interaction and cooperation in innovation systems, and their activities extend throughout innovation processes that last several years.

Innovation brokering typically comprises the following functions, to be applied in a flexible and iterative manner (Klerkx and Leeuwis, 2009):

1. **Analyzing the context and articulating demand.** The participatory assessment of problems and opportunities through quick system diagnosis identifies promising entry points (in terms of prospective markets), supportive policy, and constraining factors to be overcome. The analysis provides information to stipulate a shared vision and articulate demands for technology, knowledge, funding, and other resources.
2. **Composing networks.** Facilitate linkages among relevant actors, specifically, by scanning, scoping, filtering, and matchmaking possible partners that have complementary resources such as knowledge, technology, and funding. This also includes matching demand and supply in pluralistic advisory and research systems.
3. **Facilitating interaction.** Action planning, along with the identification of and support to those taking leadership in multistakeholder activities, has the main objective of building functioning stakeholder coalitions. Considering the different backgrounds of the actors involved, coalition building requires continuous "translation" between actors, the building of trust, establishing working procedures, fostering learning, motivating, managing conflict, and intellectual property management.

Broadly, innovation brokering can be expected to have immediate and long-term results. Direct results are expected through market innovations that arise when producers respond better to the needs of agribusiness and agribusiness operators develop a better understanding of production systems, as in the case of potatoes for the snack food industries in Kenya and Peru.

9.7. Extension System as Ecosystem Orchestrators

Extension system as Ecosystem orchestrators have to address the challenge

Box 9.3

The 1917 iTEAMS of the Government of Meghalaya can be considered as India's first ever disruptive farmer-centric, market-oriented, cloud based facilitator service that connects farmers to markets through real time agro advisories, affordable logistics and market information. The programme was established by the Department of Agriculture in collaboration with Department of Animal Husbandry, Meghalaya Agri Business Consortium, Department of Information Technology, Meghalaya Institute of Entrepreneurship, Central Agricultural University and Digital India Corporation.

The platform works purely on a business model, which is partly subsidised by the Government. The 1917 iTEAMS system is cloud based, where its core technology system is hosted remotely in servers located in Delhi. All the calls made to iTEAMS is via an Interactive Voice Response (IVR) system and linked to a database of farmers connected and registered from across the State. Callers are routed to different Incoming Call Officials (ICOs) based on their specific requirements. Those who require an Agri Response Vehicle (ARV) are directed to the Vehicle Despatch Team, those requiring advisories are routed to an ICO Level 1 Expert. If the query is not closed by the L1 Expert, it is escalated to a Level 2 Domain Expert for resolution. The ARV operators are also tracked via a GPS System and Vehicle Tracking Software developed locally, which keeps a tab on the trips, distance, time and location of the vehicles.

The project is designed with a very lean and agile model of functional, operational and staffing system. The ARVs and drivers are outsourced from transport operators who are assimilated onto the iTEAM platform. This provides the business environment of a corporate sector agency with the flexibility of a private agency along the lines of a PPP model.

Source: Field Notes of AESA, April 3, 2020.

of motivating the required partners to commit and contribute to the ecosystem. Adner (2021) identified two important risks for the feasibility of an emerging ecosystem: co-innovation risk and adoption chain risk.

Co-innovation risk stems from the fact that developing a new or substantially improved value proposition is typically associated with high risks for the individual required innovations. In the case of a business ecosystem, these individual risks multiply because of the interdependence of the different components.

Adoption Chain Risk is "the extent to which partners will need to adopt an innovation before end consumers have a chance to assess the full value proposition." Adoption Chain Risk applies the idea of a supply chain to innovation.

Ecosystem is an important design choice because it creates an indirect form of control appropriate to the complexity and dynamism of an ecosystem. It establishes the standards, rules, and processes that define an ecosystem's formal or informal constitution.

Innovation Cloud lowers the risk of failure and supports the launch of successful innovations. The Innovation Cloud Software solution is designed in such a way to help lower the risk of making wrong decisions throughout the entire innovation process. It can be ensured that only the best ideas are chosen for further development.

9.8. National Agricultural Innovation Project (NAIP)

NAIP is a major reform initiative in both science of discovery and science of delivery implemented for six years (2006-12) by the ICAR, with funding from the GoI and the World Bank. The project's development objective is "to contribute to sustainable transformation of agriculture and accelerate the collaborative development and application of agricultural innovations between public research organizations, farmers, the private sector, and other stakeholders."

NAIP seeks to accelerate the collaborative development and application of agricultural innovations involving public research institutions, farmers, private agencies and other stakeholders. ICAR has taken up the responsibility for catalysing institutional change through various initiatives in the areas of policy, strategy, governance, financial management and accountability mechanisms and a massive human resource development effort.

The project also promoted the development of three kinds of multistakeholder, multidisciplinary consortiums of public and private organizations, universities, NGOs, and others focusing on three high-priority research themes. **Market-oriented, collaborative research alliances** focus on sustainably improving the productivity, profitability, and sustainability of selected agricultural value chains. **Livelihood research alliances** focus on strategies to sustain secure rural livelihoods in about 110 disadvantaged districts, mostly in rainfed, hilly and mountainous, dryland, tribal, and coastal areas. **Basic and strategic research alliances** focus on well-defined areas of frontier science with potential applications for problems in Indian agriculture.

The underlying principle of NAIP is the formation and management of consortiums that bring research institutions together with those who use research results. The collaborative arrangements developed by these stake-holders optimize the use of research resources in an enhanced process of innovation, value addition, commercialization, and technology transfer that solves specific agricultural development problems.

NAIP has established 51 market-oriented collaborative research alliances of public, private and non-governmental organizations, farmer organisations and also international organisations. Applied research has focused on technological innovation in disadvantaged rural areas. Basic/strategic research focused on areas such as biotechnology, nanotechnology and post-harvest technology. A large portion of the research funding provided under NAIP is channelled through competitive grants to research partnerships involving public, private and NGOs.

NAIP gave public sector scientists direct experience of the challenges involved in partnering with a wide range of nonconventional partners critical for innovation.

The following organisational and institutional outcomes are noticed consequent to implementation of NAIP.

- Greater appreciation of the range of skills needed for innovation and the complementary roles of diverse partners
- The importance of broad consultations with a range of actors before conceiving project ideas and developing concept notes.
- Better understanding of how to develop large scale projects with multiple partners, which can achieve significant impact
- More frequent project reviews with partners at regular intervals
- Improved facilitation of partnerships and brokering of innovation processes
- Increased confidence of scientists to work with private and NGOs

NAIP introduced greater pluralism into agricultural research, given that 38 percent of consortium institutes (leaders and partners combined) come from outside the ICAR–state agricultural university system. Through NAIP, ICAR has promoted public-private partnerships on a large scale for the first time. The synergy created through partnerships has improved the quality and usefulness of research results.

Technology consortiums have several novel aspects. The institutional setup of the consortiums was a major innovation because it gave a market orientation (from the companies) to research on technical solutions (to the science-based institutions). This intermediate approach does not imply that research and business should change in any fundamental way; it is simply a means to strengthen the relationship between private enterprise and science.

The large change in organizing research within an AIS is the development of linkages between research and an emerging private sector. Research and bridging organizations are not well organized to effect such linkages (Larsen, Kim, and Theus 2009). In small agrarian economies, the private sector is not very prominent in rural areas

Innovations that have broad impact on the rural non- farm economy thus tend to focus on organizational innovations that improve marketing efficiency or on small-scale processing where initial market conditions exist to specialize outside the household. The development of technology for small-scale processing relies primarily on mechanical innovations that usually come either from private industry, international borrowing, or occasionally (for specialized processing) from university engineering departments or industrial research institutes (Haggblade *et al.*, 2007). The availability of specialized processing equipment depends, however, on the presence of local private capacity for manufacturing and distribution.

9.9. Need for More Effective Commercialization

Two types of diffusion and dissemination efforts are required for commercialization. Most innovations are simple and low cost (such as a pulley that locks a rope in place so that the object does not slide back, or a simple agricultural tool), and take the drudgery out of some work. The cost of diffusing such products is often higher than the cost of the product itself. But because they are simple, low cost, and easy to replicate, such products can have large social impacts on the livelihoods and quality of life of marginalized sections of society. Diffusion of these types of innovations requires the following:

1. Creating awareness of them through the media, and publicizing results of trials and demonstrations
2. Networking with NGOs and other actors with outreach to communities, building feedback loops for adoption
3. Creating a national fund to acquire rights to such technologies
4. Contributing resources to support diffusion of open source public domain technologies (social technologies), governed by the prior informed consent (PIC) of the knowledge holders, communities, and individuals.

9.10. Demand Driven Innovation

There have been significant advancements in science, its discovery, management and transfer of technology to end users. NIAP-ICAR reviewed all aspects of agricultural research and development (R&D) in India that covered research, extension, commercialization of technology, intellectual property rights, regulatory reforms, and impact of technology (Pal, 2017).

Core to the modem approach is Demand Driven Innovation that leverages participatory research to engage farmers and key value chain actors to design, develop and deliver relevant solutions -all with a sense of urgency. This approach has enabled the compression of time to deliver technology and knowledge at scale.

Modernization of research systems draws strongly on Spatial Data Integration (SDI), especially cloud computing capabilities to integrate data assets across organizations. This will help support advanced breeding programs, model priorities and track progress using geospatial analytics and apply machine-leaming to distil complex data into actionable and relevant recommendations for farmers

A focused effort is required on the "Science of Delivery" to increase rural incomes. Each state needs to take ownership, create robust implementation plans along with adequate budget and leverage Aadhaar IndiaStack and SDI to provide the technology backbone. Given the past success of the National e-Governance Division (NeGD) in supporting e-Governance reforms, the NeGD should now pivot its focus to DFI in collaboration with State e-Mission Teams.

9.11. Innovation Platforms

In practice, approaches to technical innovation within an AIS often take the form of project-based innovation platforms. Projects usually operate for a limited time; unsurprisingly they tend to focus on quick solutions to technical problems identified within the platform. This time frame orients research institutions to adapt existing knowledge and technology.

A research institute can also foster the development of innovation platforms by establishing a specialized unit or through funding arrangements such as those in Latin American research foundations or competitive grant programmes.

There is a tendency in an AIS to focus primarily on value chain approaches, as for example with the Kenya Agricultural Productivity Program (KAPP). In India, NAIP has managed the trade-off between market-driven innovation and innovation for lagging areas by creating two funding streams to support different types of innovation clusters. In lagging areas, funding supports the transaction costs involved in establishing innovation platforms and organizing partnerships. These platforms are bound by the duration of the grant, and either the problem focus has a short- term horizon, the results can be seen as a pilot for which larger-scale funding would be sought, or the funding is structured so that follow-up grants would be awarded to promising longer-term problem areas.

World Bank (2010) outlined six changes in the context of agricultural development to highlight how innovation occurs.

1. Markets, not production, increasingly drive agricultural development.
2. The production, trade, and consumption environment for agriculture and agricultural products is becoming more dynamic and evolving in unpredictable ways.
3. Knowledge, information, and technology is increasingly generated, diffused, and applied through the private sector.

4. Exponential growth in information and communications technology has transformed the ability to take advantage of knowledge developed for other purposes.
5. The knowledge structure of the agricultural sector in many countries is changing markedly.
6. Agricultural development increasingly takes place in a globalized setting.

Most discussion on India's innovation system focuses on formal research and development (R&D) efforts and the formal part of the economy. However, India is an extremely heterogeneous economy, and most of its population operates in the informal sector.

India's green revolution is the foremost example of harnessing formal creation efforts for achieving national self-sufficiency in food grains. Over the past four decades, the green revolution has significantly improved India's food security and reduced rural poverty. It has included investments in technology, largely comprising high-yielding seed varieties (initially of wheat and later rice), chemical fertilizers, and agricultural research and extension—aided by public investments in supporting infrastructure (irrigation, roads, market institutions) and price incentives that have encouraged wheat and rice production.

Although India's agriculture faces many challenges, it also has great technological opportunities. There is huge untapped potential for augmenting value chains in agriculture through crop diversification and forward and backward linkages, including post-harvest handling and processing. First, the rich diversity of agro-ecosystems is a source of sustainable growth for the sector. Second, joint ventures between public research institutes and the private sector are seen more favourably today. Such partnerships could considerably augment R&D efforts. Third, agriculture can develop value chain processing activities in rural areas to meet the changing pattern of food demand in the country and to tap international markets. Fourth, crop diversification has become a potential source of agricultural growth, creating new export and employment opportunities. Fifth, livestock, fishery, and horticulture are emerging as important sunrise sectors. Because most livestock is owned by small and marginal farmers and landless households in rural areas, the rapid growth of these sectors benefits poor households.

9.12. Institutional Innovations in Service Delivery in Agriculture

Though much of the research has emphasized technical innovations, it is equally if not more important to understand institutional innovations. Ruttan (1989) has demonstrated that institutional innovations contribute to agricultural development in a significant way. There are models of institutional innovation from a small farmer perspective which is very important to achieve

Box 9.4

In 2016, the Ministry of Agriculture, Forestry and Fisheries (MAFF) of Japan launched the Council of Industry-Academia-Government Collaboration for "The Field for Knowledge Integration and Innovation (FKII)". FKII aims to serve as a cross-sectoral platform of people, information and funds for agricultural research. The FKII is composed of three-layers:

- ☆ The Council of Industry-Academia-Government Collaboration that exchanges information among members such as producers, private companies, universities, research organizations.
- ☆ A R&D Platform that engages in collaborative research led by a designated research director.
- ☆ A Research Consortium that perform joint research.

Cross-sectoral collaboration through FKII is expected to promote the commercialisation of new technology and enhance incentives for private investment in agricultural R&D. Most of the private companies engaged in R&D related only to the merchandising or commercialisation of their own products or services. Universities and public research organisations have been playing a central role in basic and applied research, which requires a long time horizon before results are commercialised.

FKII is intended to connect each R&D stages with the participation of diverse actors of the Agricultural Innovation System. The financial support through FKII focuses on the commercialisation of research output from basic to applied, in a three-to five-year period.

The National Agriculture and Food Research Organization (NARO) is the core institute in Japan for conducting research and development on agriculture and food. NARO Bio-oriented Technology Research Advancement Institution (BRAIN) provides support for R&D that adopts a new support mechanism via a matching fund method designed to encourage collaboration with private-sector companies, as a model R&D project that uses new open innovations conducted in the FKII. A total of 17 agricultural, forestry and fisheries projects were adopted in the model projects offered from public proposals in 2016/17. They include, "development of a model plant factory system for the Asian monsoon region" and "development of an AI robot-operated greenhouse."

Source: www.knowledge.maff.go.jp/uploads/0f46f5c7b37748f264f227c4073ffd134b453ba3.pdf.

inclusive agricultural growth and development. There is a role for innovations, institutions, and institutional innovations in achieving inclusive agricultural development in a context like that of Indian agriculture.

Institutional innovations refer to change in the ways of doing things, which involves rules, norms, organisations, and organising mechanisms, as it has been recognised that institutional innovations are as important for sustainable growth and development as technological innovations (Gatzweiler, 2016).

Inputs form the backbone of Indian agriculture and they have played an enormous role in the development and success of this sector. In the recent past, there have been many experiments in the agro-input sector in terms of new distribution and marketing channels and some players have attempted to deliver total solutions to farmers including farm and allied inputs. These new channels range from marketers' own outlets to supermarkets to franchised outlets besides traditional mainstream channel of selling through distributors and dealers/retailers.

The major ones include: DCM Sriram Limited (DSL)'s Hariyali Kissan Bazaar (HKB), ITC's Choupal Sagar, Triveni Khushali Bazaar, Khushali Krishi Kendras, Champion Agro, Future's Aadhaar, and Mana Gromor of Coromandel Group. They also operate in/across different states of India. There are also agro-start-ups like Farms and Farmers and its agrobusiness arm- Green Agrevolution Pvt. Ltd. (GAPL) and Zamindara Farm Solutions (ZFS) which also attempt to achieve the same objectives for small farmers.

Singh (2018) have conducted a detailed field study on the custom rentals of farm machinery and equipment as an institutional effort supported by Punjab state, super market retailing by a local agro-supplies supermarket chain, Khushali Krishi Kendras (KKK) operated by Hydric Farm Inputs Pvt. Ltd. which is an enterprise of Rohtas Enterprises in Uttar Pradesh, and Agro-franchising for Farm Input and Market Linkage Delivery in Bihar.

Box 9.5

The Meghalaya Farmers' (Empowerment) Commission is a statutory body notified by the Government of Meghalaya in May 2019. The Commission has been envisaged as a bridge between the Government and the farmers. Empowering the farmers through collectivization and inducting technologies that are likely to add value and enhance their incomes is one of the critical objectives of the Commission. The Commission periodically advises the Government over issues of specific concern to the farmers and helps evolve policy. The organization has been responsible for introducing Buckwheat into Meghalaya and is currently engaged in expanding the Long Pepper production in the state, apart from inducting the commodity-specific technologies in Lakadong Turmeric, the globally famous product of Meghalaya. The Commission is a single-member body assisted by technical experts drawn from various fields.

9.13. Service Delivery Framework

The framework aims to guide both the design and decisions of the interventions. The basis of the framework is the simple logic that service provision needs to follow a demand and supply chain. They must interface well in order to be effective. For rural services, the demand side comprises rural communities and their organisations- the beneficiaries of the services, while supply side is more complex, consisting of the direct service providers as well as the service providing organisations and their institutional settings.

Organising the demand: An organisational demand considers the differentiation of needs and demands of various clients/social strata. It has to be inclusive. The quality of demand is an important element. Service demand can mostly be responded to in a cost effective manner if a critical mass of people share the same challenge and demand.

Experience with public service providers has shown that without effective articulation, it is unlikely that service providers will respond to specific requests, instead they will tend to remain at the 'goodwill' level of individual extension workers.

Responding to the demand: This denotes the level of the service providers. The delivery of the services need to be managed and organised to respond adequately to the articulated service demand. Service providers must be able to interpret the demand and identify the type of services appropriate for the various client's support. In a pluralistic environment, it is important to assess the performance and the quality of the services. Linkages between service providers along the platforms of the innovative system or market/value chains are critical to make the system work effectively.

Supporting the response: It describes the management of the organisational and institutional arrangement aspects of service providers. Policies and legislation regulating service provision modes and arrangements as well as finance of services need to be enabling for service providers to perform optimally. Institutional reform processes are seen as on-going adaptations propelled by learning from experiences in the field and changes in the environment.

In any intervention, the three levels need to be considered as one system.

The framework to innovate in an inclusive and integrated manner to deliver services draws on the principle of convergence, consortium, capacity building and collective action (4 "C"s) which are critical in developing strong intermediaries to address the articulated need through 4 "E"s (Efficiency, Economic gain, Equity and Environmental protection) which are the important pillars for inclusive development.

Research institutions need to work with rural entrepreneurs to accelerate the delivery of science-based solutions that can attract youth back into agriculture and treat agriculture as a business. To harness the benefits of scale in value chains, cooperation and collective action through Farmer Producer Organizations (FPOs) is needed.

Matching farmers' surpluses with market demand and enabling entrepreneurs to compress value chains that provide primary processing closer to rural communities will reduce losses, increase convenience and diversify diets for better nutrition and resilience. Most agricultural commodities have long and often inefficient value chains with a host of intermediaries. However, the era of online marketing is now influencing innovation in the agrifood sector to compress value chains and innovative business models are delivering higher prices to farmers and higher quality to consumers. This is an innovation space that is evolving quickly and will bring significant disruption into existing market mechanisms.

9.14. Extension System and Innovation for Development

The scheme for a National Innovation System for Development designed by PROCISUR in its document (PRINIDES 2010) illustrates the necessary components for a virtuous circuit leading to Innovation for Development. Its central circle is precisely the scenario where sub-components must interact, losing their traditional individual identity in order to become gears of shared innovation management for development.

Box 9.6

Since 1996, BASIX has piloted and scaled up livelihood promotion initiatives in which an integrated triad of services—financial services, agricultural, livestock and enterprise development services, and institutional development services—is delivered on a fee-for- service basis. The mix of services enabled customers (primarily the rural poor) to increase their incomes from their crops, livestock, and other enterprises. While it is widely under- stood that financial services alone are insufficient for promoting livelihoods, BASIX provides an innovative strategy for offering such integrated services in a financially sustainable manner.

BASIX delivered what it called "microcredit plus" services to clients. The "plus" components were Technical Assistance and Support Services (TASS), provided through various programmes, *viz.*, Dryland Agriculture Productivity Enhancement Programme (DAPEP), Inter Borrower Exchange Programme (IBEX), External Resource Person Advisory Programme (ERAP), Self Help Group Quality Improvement Programme (SHGQIP) and Rural Infrastructure Revival (RIR). These programs delivered services on a full grant basis or 50:50 or 75:25 cost sharing by customers and BASIX, depending on need and customers' willingness to pay.

Source: www.basixindia.com

Extension- Innovation Development (at the centre)

a) Public Policies
b) Scientific technologies
c) Production-Agribusiness
d) Rural-Territorial development (at four corners)

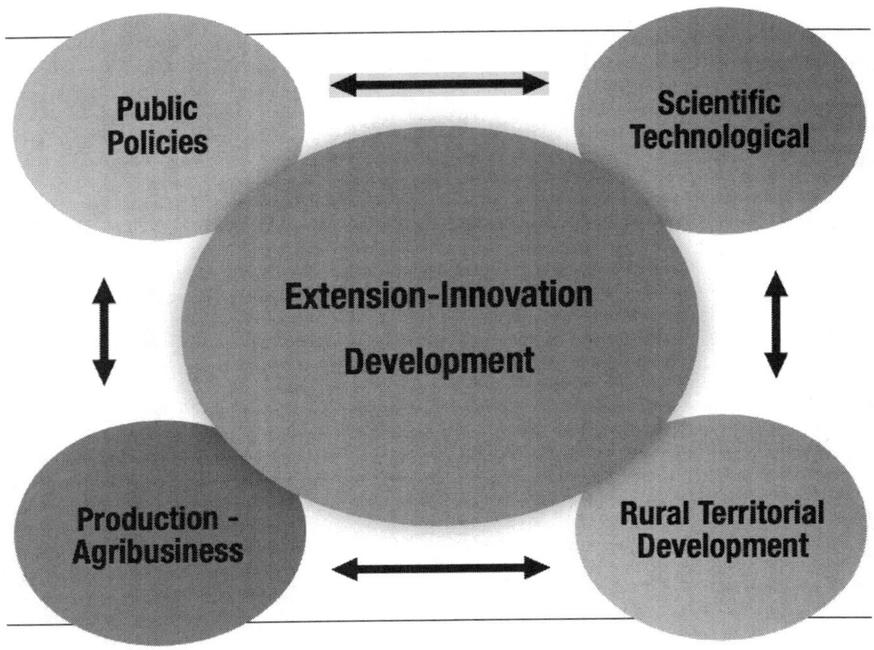

Figure 9: National Innovation System for Development.

Based on this concept of innovation, extension shares production and organisational technologies due to its joint operation with research, political system, producers and agribusiness organisations and social organisations of rural communities. Extension implies action with others, which is aimed at the dynamization of spaces where the exchange of information and knowledge boosts innovation process.

Within a context that requires co-ordinated actions between the different stages of innovation to ensure the impacts and emerging challenges, four functions of Extension are identified, that are combined according to their potential features. Considering that innovation emerges from social interaction processes, the roles of extension have to be widened and work teams must consider themselves as organisational managers, territorial complexity analysts, technological referents and, especially, as public managers of innovation processes.

1. **Organisational and Communication Managers:** There are two dimensions- institutional change and network articulation and formation. In both dimensions, management issues are vital.
2. **Territorial complexity analysts:** In a constantly changing reality, it is necessary to be capable of addressing the multiple aspects of territorial problems.
3. **Technological referents:** Different production and organisational alternatives are shared and the extension functionary must be able to arrive at the technological intervention programmes combining activities aimed at achieving products and operating in the given processes.
4. **Public managers of innovation process:** Extension teams must serve as agents that manage public policies and dynamize innovation.

9.15. Paradigm Shift Needed in Agricultural Research and Technology for Better Service Delivery

To address the new challenges and opportunities in agriculture sector, a paradigm shift is needed as below:

1. Smallholder Farmers at the Centre Stage: Recognizing that access of smallholders to technology, land, other production resources, credit, and capital is limited, a holistic pro-smallholder approach and robust policy initiatives are called for.
2. Nutrition Security to be a Key Goal of Agriculture – The extension strategy needs to shift focus on achieving comprehensive nutrition security than just food security. Decentralized research approaches will be needed for diverse ecologies, food preferences and market options, which call for region-specific prioritization of research, with due consideration to various socio-economic parameters.
3. Research, Science, Technology, Innovation (RSTI) framework would need to play a bigger role encompassing six dimensions of food security: availability, access, utilization, stability, agency, and sustainability.
4. Climate Smart Agriculture (CSA): As India is projected to be one of the most negatively impacted countries by the climate change, a paradigm shift to imbibe interdisciplinary, participatory, and international collaborative approach can address the multiple challenges caused by climate change while sustaining the desired economic growth.
5. Greening the Growth: A major paradigm shift adopting a soil, water, and biodiversity-centric approach is required to increase and sustain agriculture-food system productivity and growth that conserves

Box 9.7

India Brand Equity Foundation (IBEF) is a Trust established by the Department of Commerce, Ministry of Commerce and Industry, Government of India in 2003 with the objective of promoting and creating international awareness of the Made in India label in markets overseas and to facilitate dissemination of knowledge of Indian products and services. Towards this objective, IBEF works closely with stakeholders across government and industry.

IBEF endeavours to ensure consistent and effective messaging on the India growth story through brand building initiatives/campaigns undertaken in partnership with various stakeholders ranging from lead industry associations to export promotion councils. IBEF is the branding and communication partner for various trade exhibitions organised under the aegis of the Department of Commerce.

Some major investments and developments in agriculture as reported by IBEF in the recent past are:

- India and Brazil have signed a bilateral investment agreement, aimed at enhancing cooperation in areas of agriculture, cattle genomics, ship building, pharmaceuticals, defence production, ethanol production and oil and gas, between the countries.
- Zephyr Peacock, the India-focused private equity fund of US-based Zephyr Management, has invested an undisclosed amount in Bengaluru-based potato seeds firm Utkal Tubers India Pvt Ltd, which will be used to produce high-quality mini-tubers in a tissue culture laboratory and multiply them in its own development farms and through supervised contract farming in different regions of the country.
- Mahindra Agri Solutions Ltd (MASL), a unit of Mahindra and Mahindra Ltd, has agreed to purchase 60 per cent stake in OFD Holding BV, a Netherlands-based fruit distribution company, for Rs 36 crore, which will provide MASL access to European and Chinese markets for Indian grapes.
- Mahindra and Mahindra Ltd has acquired 35 per cent stake in a Finnish combine harvesters manufacturer and will jointly focus on the combine harvester business in Asia, Africa and Eurasian Economic Union countries.
- Mahindra and Mahindra (M and M), India's leading tractor and utility vehicle manufacturer, has entered into pulses retailing under the brand 'NuPro'. The company plans to foray into e-retailing and sale of dairy products.

- ☆ Fertiliser cooperative IFFCO launched a joint venture with Japanese firm Mitsubishi Corp for manufacturing agrochemicals in India.
- ☆ Acumen, a not-for-profit global venture fund, has invested Rs 11 crore in Sahayog Dairy, an integrated entity in the segment, based at Harda district in Madhya Pradesh.
- ☆ Rabo Equity Advisors, the private equity arm of Netherlands-based Rabo Group, raised US$ 100 million for the first close of its second fund, India Agri Business Fund II. The fund plans to invest US$ 15–17 million in 10 to 12 companies.
- ☆ Oman India Joint Investment Fund (OIJIF), a joint venture (JV) between the State Bank of India (SBI) and State General Reserve Fund (SGRF), invested Rs 95 crore in GSP Crop Science, a Gujarat-based agrochemicals company.

Source: IBEF Report, December, 2017.

natural resources and mitigates climate change, moving from Green to Evergreen Revolution to Evergreen Economy

6. From Subsistence to Commercial Agriculture: Good post-production technologies, robust entrepreneurship initiatives, and establishment of agro-industries in rural areas should foster economic and employment security and attract youth in agriculture-food systems, thus harvesting India's huge demographic dividend and achieving agrarian transformation.
7. From Green Revolution to Gene Revolution: Genomics and gene editing must be adopted as preferred technologies for precision breeding. However, to do so, a science-based policy and well defined guidelines need to be placed on priority. In the new era of genomics, phenomics, proteomics and other omics, molecular breeding, the availability of high quality reference genomes of crop plants accelerated the discovery of genes, quantitative trait locus (QTLs) and DNA markers linked to the traits of agronomic importance. This is now being routinely applied in MAS of crop varieties, including those of horticultural species, for increased selection efficiency.
8. Digital Solutions and Artificial Intelligence (AI) for Evergreen Revolution: Using Big Data Analytics, we can create decision support systems at various levels, including weather forecasting and efficient management of water, pest, and nutrient. Linked with satellite imagery, this can also help in future predictions of produce and price
9. Towards Precision Agriculture: Precision agriculture, which exploits modern tools, technologies and innovations, including genetically

enriched seeds, nanotechnology, artificial intelligence (AI), drones, sensors, robots *etc.* is the way forward to achieve environmentally sustainable Evergreen Revolution.

The rapid evolution of molecular biology and information technology to integrate across disciplines and sectors will play a critical role in India to unlock innovation and realize impact in farmers' fields. While the agricultural research and development system in India has responded to these developments, more can and must be realized by using modern tools (cloud computing, artificial intelligence, mobile, remote sensing, and systems research) to support agrarian reforms to make agriculture a commercially attractive and sustainable enterprise for current and future generations.

Under Digital India interventions of the Government Agriculture 2.0 (Digital Agriculture) has been given the prominent place for improving overall development of the Country through increasing the agricultural output by ensuring availability of information on various agriculture and allied sectors activities. Agriculture 2.0 (Digital Agriculture) directly falls under Pillar No. 5 of Digital India, *i.e.* eKranti - Electronic Delivery of Services and broadly caters to other pillars as well, like e-Governance: Reforming Government through Technology, Information for All and Early Harvest Programmes.

9.16. Inclusive Innovation in Agriculture

A new approach to innovation that explicitly benefits the least well-off in terms of process and outcomes, and which can contribute to development that is more inclusive has been suggested (Heeks *et al.*, 2014). The approach cannot be considered as entirely new, since pro-poor innovation, grassroots innovation, below-the-radar innovation and related concepts have been well documented since the 1990s.

Unlike the term "participatory innovation," which has been in use among agricultural and development practitioners for nearly four decades, "inclusive innovation" has entered into use in the past decade and still lacks a consensus definition (Heeks *et al.*, 2014, Pansera and Owen, 2018). The most widely referenced framing refers to inclusivity in terms of who benefits from innovations and innovation processes, and is focused specifically on the inclusion of poor and marginalized populations, sometimes referred to as individuals at the base of the pyramid (BoP) (Heeks *et al.*, 2014). Inclusivity in this regard can refer to the extent to which innovation processes address the challenges and concerns of poor and marginalized communities and the extent to which these communities benefit from the resulting innovations, either through direct uptake or because the innovations contribute to including them in systems, markets, or benefit streams from which they were previously excluded (George *et al.*, 2012).

A more comprehensive framing of inclusive innovation has been developed by Schillo and Robinson (2017), who contend that innovation needs to be inclusive in terms of people, activities, outcomes, and governance. They consider four dimensions to determine if an innovation process is inclusive: 1) who participates in the innovation process; 2) what kinds of activities the process entails (activities that are explicitly participatory and that enable the meaningful participation of diverse stakeholder groups), 3) the results or outcomes of the innovation process and who benefits from these; and finally, 4) how and by whom the innovation system itself is governed. This framing is generally consistent with the "ladder of inclusive innovation" proposed by Heeks *et al.* (2014), which considers six levels of inclusion ranging from the intended impact (at the lowest level of inclusion) through actual impact, to process, structure, and "post-structure," referring to the inclusivity of the frame of knowledge and discourse within which innovation is created.

To enable inclusive innovation, First, we need to shift our emphasis from looking at only technical innovations and should focus more on the role of institutional innovations to promote inclusive innovation. Second, we need to recognize the role of the rural poor as co-learners, active learners, and as repositories of much useful knowledge rather than looking at them as just beneficiaries or users.

Box 9.8

The K-Rail Silver Line is an ambitious project of the Kerala state government that aims to cut the travel time between the north and south ends of the state to four hours from 10-12 hours. It is a proposed government-funded high-speed rail line that would connect Kasaragod in the north with Thiruvananthapuram in the south.

The project has stirred a massive row in Kerala, with experts reportedly saying that the project will displace thousands of families and will also damage the environment. Over 20,000 families have to be displaced for the project and the state has to find out a whopping amount for its implementation. As per NITI Aayog, it might cost Rs 1.24 lakh crore when it nears completion in 2025.

Kerala is a highly urbanized and mixed-growth ecosystem with a larger metropolitan. The state could bridge the divide between rural and urban systems because of the remittance economy and migration from the interiors. In an already-fractured economy, the state must prioritize higher-quality education, job creation opportunities, and environmental protection. Regrettably, the proposed K-Rail fails to meet any of these criteria. Instead, it goes against Kerala's tradition of decentralization while encouraging involuntary movements through inducement.

Source: http://www.timesnownews.com>india

"Innovation" is related to "inclusion" in the following ways:

First, "inclusive innovation" can provide solutions for reducing gaps in living standards between the richest and poorest groups in society. Such innovations typically consist of producing cheaper (often simplified) versions of existing often sophisticated products for purchase by lower-income groups

Second, some innovations facilitate grassroots entrepreneurship and could help integrate previously marginalised groups into circuits of economic activities. It often involves either the exploitation of traditional knowledge (agriculture, craftsmanship) or an adapted use of modern technology that most people can afford.

Third, firms do not engage in innovation to the same extent. Differences among firms' innovation activities and use of new technologies translate into substantial productivity gaps.

9.17. Responsible Innovation in Agriculture

Responsible Innovation (RI) refers to a process of innovation that meets certain procedural norms like accountability, inclusiveness, due care and transparency to stakeholders and to the society. This means a responsible approach towards innovation which involves creating change having positive impacts on society and the environment.

RI has become a prominent topic in thinking about ethical and social aspects of innovation and new technology in the past few years. But it goes beyond consideration of ethics, public engagement, risk and regulation. The public expects the researchers to act responsibly, and RI creates spaces and processes to explore innovation and its consequences in an open, inclusive and timely way. It brings together researchers from different fields of study including technology assessment, science communication and ethics of technology. Responsible innovation encourages us to think of what technologies are for, who they are serving and who is driving the process (Crossley, 2018).

Responsible Research and Innovation (RRI) is an emerging field in the European research and innovation (R and I) policy context that aims to balance economic, socio-cultural and environmental aspects in innovation processes. The concept of RRI calls for ethical reflection on the nature, scope and applicability of responsibility of innovation and innovation practices in general, and the way social and ethical issues can be applied and addressed in agriculture. Incremental innovation has always been at the centre of agriculture for many years. Incremental innovation does not create new concepts or products, as it focuses only on marginal improvements to what already exists. However, in the last few decades, agricultural research has led to many radical innovations in agriculture such as hybrid seeds, genetically modified crops, nano products, *etc.* These radical innovations have led to much criticism from a section of the society, particularly environmental groups and animal rights activists. Though agricultural production today faces several challenges including climate and market, there is a common understanding that the environment and animal welfare need to be protected. The farmers are urged to reorganize their practices in a responsible way under various labels such as GAP, 'organic', safe to eat, healthy food, *etc.*

In agriculture, the concept of RI has not been widely considered, although it has much relevance, particularly in the context of smart farming emerging as the latest innovation in agriculture. The specificities of smart farming revolution create challenges for responsible innovation frameworks that have largely been developed in other fields, and have not been tested in practice in the agricultural context.

There are two recent reports (Eastwood *et al.*, 2017) on RI in smart dairying and Bronson (2018) on smart farming), which are found useful and relevant and briefly explained. The widespread use of smart dairy farming technologies, including robotic milking was found to involve varied outcomes for animals, people and the environment. Eastwood *et al.* conducted the study using anticipation-inclusion-reflexivity-responsiveness (AIRR) framework, identified as the key dimensions of responsible innovation.

Anticipation: To enhance anticipation in science and innovation governance, actors (scientists, practitioners, technology developers, policy makers) shall use processes to identify and minimize unintended consequences of future innovations. Potential indicators of anticipatory processes include the use of foresight exercises, horizon scanning, and scenario-building techniques.

Inclusion: This relates to broadening the debates around innovation from top-down governance mechanisms to bottom-up approach and the inclusion of all the stakeholders. The inclusion of stakeholder perspectives in technology development is suggested as a method for improving stakeholders' trust in the innovation process. Techniques to facilitate inclusion include citizen panels, focus groups, and user-centred design.

Reflexivity: Being more reflexive involves researchers assessing their own motivations and assumptions (*e.g.* social, ethical, and political norms and values) and acknowledging the perspectives of other actors (*e.g.* the public, companies, *etc.*) on particular issues. Certification and standardisation and codes of conduct have been identified as ways for public and private institutions to communicate their norms.

Responsiveness: Responding to societal needs requires the ability to change direction in the innovation process in light of emerging knowledge and perspectives. Since societal challenges, perspectives, and norms are likely to change over time, responsible innovation also needs to have the capacity to change direction. Open access to research processes and results, along with declarations of conflicts of interest, are potential responsive approaches.

The study identified changes to the nature of farming and human-animal relations as socio-ethical dilemmas, potentially leading to farmers and the society rejecting smart farming technology based applications. By applying the AIRR framework to a case study of smart dairying in New Zealand, Eastwood *et al.*, also identified lessons for the application of smart farming technologies to farmers in the future.

Proposed indicators of responsible innovation activities in smart dairy (Eastwood *et al.*, 2017)

Indicator	*Description*	*Potential Activities*
Anticipation 1. Foresight exercises	Future scanning activities	Technology use surveys, Assessing farmer perceptionof technologies
2. Scenario building of smart dairy farms	Imagine potential and negative futures	Visioning of smart dairyfarms
Inclusion 3. Involvement of relevant actors	A range of end-users and citizens are involved	Seeking critical feedback
4. Private sector engagement	Private companies are included as partners in publicly funded research	Private companies co-fund research projects
5. Encouraging transformative mutual learning	Mutual learning	User-centred design; Open innovation
Reflexivity 6. Reflexive guidance	Reflection on values around development and use of technology	Reflective monitoring
7. Structures guide second order reflexivity	Use of codes of conduct and standards	Creation of best practices guidelines
Responsiveness 8. Potential to adapt changes	Change direction based on stakeholder feedback	Mid project revisions; Milestones
9. Open research process and access to research data	Transparent smart farming design	Open data exchange, open access to data points

Similarly, Bronson (2018) studied smart farming innovation in Canada and portrayed the scenario of John Deere fitting each of its 'precision' tractors with sensors that collect data about soil and crop condition. The software used in John Deere's tractor is proprietary and the data it collects are not openly accessible. Farmers have to subscribe and pay for access to information it generates from aggregated datasets, which is accessible using a 'My Farm Manager' mobile application. The proponents of precision tractors suggest that the tools introduce incredible business efficiency into farming.

It was observed that corporates recognize the economic potential in smart innovations applied to agricultural production. In 2013, Monsanto after purchasing the digital tool developer 'Climate Corporation', under a platform

Box 9.9

The three new Farm Acts legislated by the Government of India can be treated as a case of lack of responsibility on the part of the Government, which resulted in withdrawal of the Acts. The Acts have been widely acclaimed at home and abroad as historical and long overdue. However, some experts, states, and stakeholders, including farmers, have been protesting against them and seeking their withdrawal. The Government was forced to withdraw the Acts due to lack of responsibility in educating the farmers and convincing them about the advantages of the Acts. The farmers were carried away by the political parties, resulting in lapse of three potential Acts.

Farmers' Produce Trade and Commerce (Promotion and Facilitation) Act 2020 (FPTC Act) - The best benefit from the new Act will accrue when APMC mandis and private channels coexist and compete. This can be ensured by the states by nurturing APMC mandis as infrastructure service for the farmers, like other government facilities such as hospitals, schools, roads and parks, rather than using them for generating revenue for the government and middlemen

Farmers (Empowerment and Protection) Agreement on Price Assurance and Farm Services Act, 2020 - The Act empowers the farmers for direct marketing reducing the number of intermediaries. There is provision of 'farm services' in the act, which is used by the opponents of the act to conclude that the Act is for promoting contract farming. But the real fact is that this provision is included for strengthening FPO activities.

Essential Commodities (Amendment) Act 2020- The Essential Commodities Act has been modified for agriculture and food produces including cereals, pulses, potato, onion, edible oilseeds and oils.

called Integrated Field Systems (IFS) released in 2014, offers farmers a suite of digital tools for collecting and analysing farm data. Monsanto use farm-led data to promote its proprietary chemicals and farmers' data to drive its R&D. Here, whether the individual farmer benefits commensurately with the corporate is an ethical question.

Bronson indicated that smart farming innovation in Canada has a blind spot in relation to the needs and concerns of small and medium sized labour intense farms. Smart innovations work to the advantage of already powerful players in the food system in Canada, which is a market reality, but is simultaneously a democratic problem. Technological equity and broad social progress has to be secured through careful and ethical decisions taken by key players in the innovation ecosystem. The inclusion of right holders in broad, value based inquiries related to technology development is suggested as a solution for improving outcomes and also for improving trust in the innovation process.

There is an indication that innovation in big data and machine intelligence may enable by their very design, consolidation of power among the agribusiness companies. For instance, Monsanto's Field Support, a platform for integrating farm level big data, makes use of a proprietary algorithm to arrive at a 'prescription', which is completely opaque, protected as corporate intellectual property and there is lack of transparency around the profit generating uses of Field Support big data.

Though corporates have a legal mandate to maximize profit, the Government has a democratic mandate to ensure that the technologies it helps to develop contribute to the goal of society as a whole and not just the interests of the already rich and powerful classes. Value based questioning ought to be foregrounded, as a normative matter, in inclusive process of decision making related to smart agriculture.

REFERENCES

Adner, Ron (2021) Winning the Right Game: How to Disrupt, Defend and Deliver in a changing world, The MIT Press, Cambridge

Anderson, J. R. and Feder, G. (2007) "Agricultural Extension," in R. E. Evenson and P. Pingali (eds.), Handbook of Agricultural Economics, 2007, Vol. 3, Elsevier, pp 2343-2378

Bronson, K (2018) Smart farming: including rights holders for responsible agricultural innovation, *Tech Innov. Manage. Rev*, 8(2), 7-14 doi:10.22215/timreview/1135

Crossley, D (2018) Big Questions and Radical change Available online at: https://www.foodethicscouncil.org/uploads/For per cent 20whom per cent 20- per cent 20questioning per cent 20the per cent 20food per cent 20and per cent 20farming per cent 20research per cent 20agenda_FINAL_1.pdf

Eastwood, C., Klerkx, l., Ayre, M., and Dela Rue, B (2017) Managing socio-cultural challenges in the development of smart farming: from a fragmented to a comprehensive approach for responsible research and innovation, *J.Agric.Environ.Ethics*, 32, 741-768 (2019)

David Sunding and David Zilberman (2001) The agricultural innovation process: Research and technology adoption in a changing agricultural sector in Hand book of Agricultural Economics, Vol 1, Part A, 2001 p 207-261

Gonzalez Asis, Maria; Woolcock, Michael. (2015) Operationalizing the Science of Delivery Agenda to Enhance Development Results. Washington, DC: World Bank.

Haggblade, S., Hazell P B R and Reardon, T (2007) Transferring the Rural Non-farm Economy: Opportunities and Treats in the Developing World, IFPRI Brief, 58

Inter American Institute for Co-operation on Agriculture (2014) Innovation in agriculture: a key process for sustainable development, Institutional position paper, San Jose, May 2014

India Brand Equity Foundation (2017) Agriculture in India: Information About Indian Agriculture and Its Importance- Report, https://www.ibef.org>Archives>Industry

Klerkx L., van Mierlo B., Leeuwis C. (2012) Evolution of systems approaches to agricultural innovation: concepts, analysis and interventions. In: Farming Systems Research into the XXIst Century: The New Dynamic (I. Darnhofer, D. Gibbon, B. Dedieu, eds). Springer Science, Dordrecht.

Klerkx, L., Hall, A J., and Leeuwis, C (2009) Strengthening Agricultural Innovation Capacity: Are Innovation Brokers the Answer? International Journal of Agricultural Resources, Governance and Ecology, 8 (5)

Larsen, Kurt; Kim, Ronald; Theus, Florian. (2009) Agribusiness and Innovation Systems in Africa. Washington DC: World Bank.

Pal Suresh (2017) Agricultural R&D Policy in India: The Funding, Institutions and Impact. ICAR National Institute of Agricultural Economics and Policy Research, New Delhi.

PROCISUR The role of Rural Extension in Innovation Management, The Cooperative Program for Technological Development in Agrifood and Agroindustry in the Southern Cone -PROCISUR

Wani SP, J Rockstromand B Venkateswarlu (2011) New Paradigm to Unlock the Potential of Rainfed Agriculture in the Semiarid Tropics. Pages 419- 470zn World Soil Resources and Food Security (Lai R and Steward BA, eds.)

Wani SP, Bergvinson D, Raju KY, Gaur PM and Varshney RK. (2016) Mission India for Transforming Agriculture (MITrA) Research Report IDC-4, ICRISAT, Patancheru, Hyderabad. 15pp

World Bank (2003) World Development Report 2004: Making Services work for poor people, Washington, DC: World Bank

World Bank (2007) World Development Report 2008: Agriculture for Development, Washington, DC: World Bank

World Bank. (2010) Innovation Policy: A Guide for Developing Countries, Washington D C, World Bank.

World Bank (2015) World Development Report 2015: Mind, Society and Behaviour, Washington, DC: World Bank

Chapter 10
Epilogue

Innovation acts as a catalyst in transforming agriculture to adapt and respond to various challenges and crises. Innovations in agriculture provide both opportunities as well as challenges. The farmers are always in the look-out for various innovative technologies, which are expected to improve productivity, resource use efficiency and sustainability.

Farming is a complex, unpredictable and individual operation. There are increasing pressures from climate change, soil erosion and biodiversity loss on one side and from consumers' changing tastes in food and concerns relating to quality on the other. The pests and diseases continue to pose their own challenges. While modern agriculture provides a large number of solutions, the outcome is not always the same because each farm is unique: different landscapes, soils, available technology and potential yields.

Innovation in agriculture cuts across all dimensions of the production cycle and along the entire value chain - from crop, fishery, forestry or livestock production to the management of inputs to market access. FAO of the UN plays a key role in promoting the importance of innovation in agriculture to increase food security, sustained development and promote rural development. FAO also harnesses the power of digital technologies to pilot, accelerate and scale innovative ideas with high potential for impact in food and agriculture, transforming digital solutions and services into global public goods.

10.1. Innovation Process

Agricultural research has often been confused with innovation, though there are differences between them. Research is concerned with the production of new knowledge, which may or may not be used in practice. Innovation, on the other hand, is concerned with the processes of change in the production and marketing of goods and services – changes that may or may not be driven by research. In the context of agricultural development, innovation has been defined as "the process by which individuals or organizations master and implement the design and production of goods and services that are new to them, irrespective of whether they are new to their competitors, their country, or the world" (World Bank, 2012).

At the farm level, many innovations are "process innovations" that improve production techniques, such as high density planting, intercropping, *etc.* "Product innovations" are created by downstream industries, and include new and improved products, such as new pesticide formulations, new machineries, *etc.* "Marketing and organisational innovations" are also increasingly important throughout the supply chain. FPOs have emerged as an important organisational innovation.

India's green revolution is the foremost example of harnessing formal R&D efforts for achieving national self-sufficiency in food grains. Over the past few decades, the green revolution has significantly improved India's food security and reduced rural poverty. Although India's agriculture faces many challenges, it also has great technological opportunities. There is huge untapped potential for augmenting value chains in agriculture through crop diversification and forward and backward linkages, including post-harvest handling and processing. First, the rich diversity of agro-ecosystems is a source of sustainable growth for the sector. Second, joint ventures between public research institutes and the private sector are experiencing more favour today. Such partnerships could considerably augment R&D efforts. Third, agriculture can develop value chain processing activities in rural areas to meet the changing pattern of food demand in the country and to tap international markets. Fourth, crop diversification has become a potential source of agricultural growth, creating new export and employment opportunities. Fifth, livestock, fishery, and horticulture are emerging as important sunrise sectors. (Utz and Dahlman, 2007).

For agricultural research to benefit the rural poor, it needs to complement other efforts that improve the policy environment, alleviate resource constraints, and build local capacity for responding to changing technological and economic challenges and opportunities.

Since the 1970s, international agricultural research centres have worked with national and regional partners to stimulate agricultural innovation and growth. However, it was observed in many cases, the benefits derived by smallholders have been constrained by the farmers' limited opportunities to market their products. In an attempt to expand the benefits of agricultural research and development for smallholders, since 2000, many organizations have experimented with novel approaches for promoting innovation on small farms related to marketing, including the concept of value chain development (VCD).

The institutional arrangements and standard operating procedures of agricultural R&D organizations have tended to lag behind the evolution of thinking on innovation processes and systems (Hall, 2009). The agricultural innovation system approach emphasizes the collective nature of innovation and stresses that innovation is a co-evolutionary process. These insights are increasingly orienting interventions that focus on setting up multistakeholder initiatives, such as innovation platforms and networks to enhance agricultural innovation. Several recent studies have addressed the issues of platform organization, but there has been limited analysis of how platforms shape innovation processes (Devaux *et al.*, 2018).

Innovations in R&D approaches are an important way to foster innovation processes in the productive sector, benefitting smallholders and other economic actors. Several studies indicate how the participation of research organizations in multistakeholder platforms and acting as innovation brokers has improved the linkages between researchers and other service providers and value-chain actors. This has contributed both to innovations in the productive sector and to improving the focus of applied research on the challenges and opportunities identified by value chain actors.

In India, the new Science, Technology and Innovation Policy of 2020 aims at building a nurtured ecosystem that permits research and innovation on the part of both individuals and organisations. The policy promotes technology seif-reliance and indigenization to achieve the larger goal of 'Atmanirbhar Bharat'. The new policy has a chapter on 'Innovation and Entrepreneurship' which aims at creating an enabling ecosystem for seeding, sustenance and growth of the S&T entrepreneurship in India. S&T enabled entrepreneurship and technology driven innovation focused enterprises is critical for creating, shaping and sustaining agricultural development of the country and delivering the benefits to the farming community.

There are three clusters of enabling factors for agricultural innovation that require proper attention and investment in most developing countries, including India. They are: (1) Innovation policy and corresponding governance structures to strengthen the broader framework for agricultural innovation policies; (2) Regulatory frameworks that stimulate innovation directly such as IPRs or indirectly such as standards that stimulate trade or steer innovation towards certain preferred outcomes (safer food); and (3) Accompanying agricultural investments in rural credit, infrastructure, and markets. Innovation policy is a new domain, and in most countries the governance structure for innovation is only starting to emerge.

In India, the ICAR has played a pioneering role in ushering Green Revolution and subsequent developments in agriculture in India through its research and technology development that has enabled the country to ensure food security of the country. ICAR's three tier system of technology commercialization process has Intellectual Property and Technology Management (IP and TM) Unit at the HQ as the top tier, Zonal technology Management Centres (ZTMCs) as the middle tier and Institute Technology Management Units (ITMUs) at the bottom level for effective technology transfer through licensing.

National Agricultural Innovation Project (NAIP) is a major reform initiative implemented by the ICAR, with funding from the GoI and the World Bank. The project's development objective is "to contribute to sustainable transformation of agriculture and accelerate the collaborative development and application of agricultural innovations between public research organizations, farmers, the private sector, and other stakeholders."

NAIP has resulted in accelerating the collaborative development and application of agricultural innovations involving public research institutions, farmers, private agencies and other stakeholders. ICAR has taken up the responsibility for catalysing institutional change through various initiatives in the areas of policy, strategy, governance, financial management and accountability mechanisms and a massive human resource development effort. The importance of Agricultural Value Chain (AVC) gained prominence through the NAIP research. Value chains act as an important catalyst in improving farmers' income by strengthening the backward and forward linkages of agriculture. Various innovative strategies are being made to organize farmers and facilitate their access to markets, finances, inputs and technologies.

10.2. Innovations in Agriculture

The innovations in the field of agriculture is explained in this book under three phases: Diffusion of technological innovations, Adoption of disruptive innovations and Emergence of innovations for degrowth.

Diffusion of Innovation (DOI) theory proposed by Everett M Rogers dominated the research in Agricultural Extension all over the world for almost half a century, barring certain criticisms posed. Some Extension researchers have indicated that technology adoption in agriculture may not always follow the classical adopter category- wise progression as indicated in the DOI theory. An important challenge to the DOI theory comes from Geoffrey Moore who wrote the book 'Crossing the chasm' (2002). He suggests that the adoption curve is not continuous for discontinuous technologies, where adopters are required to substantially change their behaviour or to modify other products and services they use.

The technology adoption lifecycle appears as a bell curve, sorting the adopters into five categories- innovators, early adopters, early majority, late majority and laggards. These categories seem to show a neat progression. However, in the technology adoption lifecycle, there is often a large gap between early adopters and the early majority. This gap is referred to as the 'chasm'. According to Moore, 'Crossing the chasm' means being ready to meet the needs of the next round of new customers. By being aware of the upcoming hurdle and creating a strategy for getting around it, the readiness to meet the needs of the pragmatic majority is assured leading to the improvement in the chances of success.

The field of Extension is now in a Digital disruption stage, which follows an understandable pattern. The starting point for the leaders in agriculture is to understand where in this pattern their organization is positioned and why that is the case. A public extension professional today may be happy serving farmers with the linear flow of knowledge and services (mostly it is because of their position in hierarchy, rather than the quality of services they provide), sooner or later new players will disrupt the linear flow of services and will try to provide retail like experiences. This has already occurred in the public extension service with the entry of private extension service providers.

Disruption does not happen only because of digital technologies, but through farmers centric approaches that offer retail like extension services (advisory and supply chain coordination) specifically designed to meet and exceed expectations of farmers.

- ☆ Big data and predictive analytics at Extension and Advisory Services (EAS) level can help farmers to access and apply crop choice, market recommendations, pest modelling, soil test value, and crop yield predictions, as well as nutrient management- all across varying field conditions.
- ☆ Better market price realization is difficult to achieve by simply establishing eNAM kind of online platforms. Real disruption is required in the way farmers market their produce.
- ☆ Knowledge Transfer through personalised advisory and multiple delivery channels have been tried out by digital start-ups
- ☆ Aggregation model all along the value chain brings together small farmers for accessing numerous services (mechanization, bulking, input buying, markets *etc.*). A careful analysis of several start-ups reveals ways and means achieving this using digital technology.
- ☆ Financial Inclusion of small and marginal farmers includes credit and insurance. Quite a few models are available across Asia and Africa on this front. Digital financial services (DFS) are fundamentally about saving money, accessing credit and insurance, and performing transactions via digital channels like eNAM.

In terms of digital disruption, public sector is uniquely positioned in terms of scale, reach and data, if only proper strategies are adopted. To realise digital disruption in public EAS, there is need to create avenues for redesigning the extension processes (work on the frameworks not on the guidelines), stimulate new thinking (capturing innovations and startups within the system) and include crowd sourced extension innovations (allowing partnerships and local redesigning to certain extent).

Deloitte, one of the largest global professional services networks, has published a book titled 'The Transformation Myth: Leading Your Organisation Into An Uncertain Future' in an effort to debunk the common misconceptions preventing businesses from successfully navigating digital transformation and attaining a top position. This book is aimed at using disruption as an opportunity for innovation and broaden their understanding of how technology enables essential capabilities for survival.

In the context of the crisis-filled 2020, the book indicates that companies that use disruption as an opportunity for innovation emerge from it stronger. Companies that merely attempt to "weather the storm" until things go back to normal (or the next normal) miss an opportunity to thrive.

Drawing on five years of research into digital disruption, including a series of interviews with business leaders conducted during the COVID-19 crisis, the book provides a framework for understanding disruption and tools for navigating it. The Transformation Myth also offers insight into the leadership traits, business principles, technological infrastructure, and organizational building blocks essential for adapting to disruption, with examples from leading organizations.

The important messages from the book are:

- ✰ Technology isn't an end in itself. Rather, it enables the capabilities essential for surviving an uncertain future: nimbleness, scalability, stability, and optionality.
- ✰ The innovation ecosystem should promote wider diffusion of new technologies. The century-old patent system should be reviewed in light of the realities of today.
- ✰ Public investment in R&D should be revived to strengthen support for innovation that serves broader economic and social goals rather than interests of narrow group of investors
- ✰ The Small Business Innovation Research and the Small Business Technology Transfer programmes should be strengthened and their effectiveness enhanced by shifting more resources to early stage awards to small and young firms.

In her book 'Regenerative Agriculture: Policies and practices for sustainability and self reliance', Pretty (1985) explains that 'most successes are still only small scale... only a few thousand communities throughout the world have benefitted'. She also argues that mainstream policies that do not reflect the long term social and economic costs of resource use, act as powerful disincentives to change towards more sustainable forms of rural activities.

Convivial innovation is characterized by the need to slow down the process of excessive use of natural resources affecting sustainability. User friendly and nature friendly innovations such as natural farming, eco-responsible construction, home-made clothing, *etc.* enables the rediscovery and revaluation of the traditional knowledge.

Herman *et al.* (2020) introduced four lenses that contribute to better understanding complex innovation dynamics: (a) social dynamics and information transfer, (b) contextual costs and benefits, (c) experience and risk aversion, and (d) practice adaptation. Investments should build on existing knowledge and farming systems including a focus on the dynamic decision process to support the 'scaling up, scaling out and scaling deep' agenda for sustainable agricultural innovations.

Agroecology is a dynamic concept that has gained prominence in scientific, agricultural and political discourse in recent years. It is increasingly promoted as being able to contribute to transforming food systems by applying ecological principles to agriculture and ensuring a regenerative use of natural resources and ecosystem services while also addressing the need for socially equitable food systems within which people can exercise choice over what they eat and how and where it is produced.

Agroecology is a transdisciplinary science, combining different scientific disciplines to seek solutions to real world problems, working in partnership with multiple stakeholders, considering their local knowledge and cultural values, in a reflective and iterative way that fosters co-learning among researchers and practitioners.

Agroecological practices harness, maintain and enhance biological and ecological processes in agricultural production, in order to reduce the use of purchased inputs that include fossil fuels and agrochemicals and to create more diverse, resilient and productive agroecosystems.

Conceptually, there is a need for a radical shift, beyond incremental improvements and technological fixes, to new kinds of socio-technical systems, often called 'sustainability transitions', referred to as degrowth innovations.

10.3 Innovative Methodology for Extension Service Delivery

Prasad (2017) has developed a concept note on an innovative methodology of Agricultural Extension titled 'Facilitation and Convergence (F and C) system' of Extension. After T and V system of Agricultural Extension, there is no system of extension successfully implemented in India, though many approaches are tried like group farming, market-led extension, cyber extension, *etc.* There is an imperative need to develop a **demand side governance** mechanism for effective delivery of services. The call for an innovative extension service delivery is made at a time when the existing methodologies have not succeeded in providing the right type of services needed by the client system. The concept on Facilitation and Convergence (F and C) system of Extension is suggested as an attempt in this direction to fill the existing gap.

Facilitation refers to the process of developing an enabling environment for farmers to take proactive role in participating and benefitting from different extension services, *viz.*, extension plus, empowerment extension, partnership extension, cyber extension and livelihood extension.

Extension personnel have to engage in the following three categories of facilitation as below:

- ✰ **Technological facilitation**- for productivity and profitability enhancement- Better farming skills
- ✰ **Process facilitation**- Facilitation in the contents and quality of extension service is to be attempted with focus on competitiveness, social sensitivity, environmental concern and sustainability- IT led extension, market led extension
- ✰ **Empowerment facilitation**- for new skills and competence for improved livelihoods- Developing chain partnerships and ownership over the chain- *e.g.*, Mobilizing farmers and organising FPOs

Capacity building programmes have to be organised with the help of domain experts to assist the farmers to acquire practical skills to improve their access to resources and to create space for their actions by playing supporting and enabling roles. Capacity building programmes may be designed to comprise of three inter-related actions- a) development of intellectual skills and a knowledge base; b) gaining access to resources through mobilisation of own efforts and through assertion of rights and claims, to develop the productive base for livelihood improvement, and c) development of capacities for self-organisation and management of development actions. Skill development efforts of the Government need to be fully tapped.

Convergence refers to the process of developing and nurturing structural and functional relations with various organisations. The process of adapting the needs and requirements of the farmers with the existing schemes of the government in a convergent mode will enable better planning and result in better support system services for the farmers.

10.4. Suggestions for Innovation Research in Agricultural Extension

Based on analysis of various concepts and principles included in the book, the following suggestions are made for organising more effective innovation research in the field of Agricultural Extension.

- ☆ Innovation as a concept has a profound role and influence in technology transfer and related extension services. Technological innovations in agriculture along with innovations in the delivery of extension services need to be analysed more rigorously for assessing the impact and process of scaling up.
- ☆ Research on innovation in the field of Extension education is mostly confined to communication of innovations, diffusion of innovations, attributes of innovation, agricultural innovation system, *etc.* There is need for more systematic studies on the communication and diffusion of innovations in the context of group approaches, influence of social media, *etc.* More of action research in this area needs to be encouraged.
- ☆ Innovation involves more than research and development and also entails the functioning of the market system. In this respect, the field level studies on Agricultural Value Chain assume significance.
- ☆ There are multiple meanings for the concept of innovation within different paradigms predominant in different times under various economic, political, social and cultural contexts, emerging from different disciplines. However, the various dimensions of innovation are not fully captured and utilized in Extension research, which is a critical gap. This needs immediate attention of extension researchers.
- ☆ Innovation and knowledge systems in agriculture in their broader sense have to be understood as the articulated group of agents, networks and or organizations linked to development, science, technology and extension system, networks and organizations of producers, community organizations and different levels of governance system. Studies on innovation ecosystem and governance in the field of agricultural innovations are largely needed.
- ☆ More focused and exploratory research on the various emerging dimensions of innovation is strongly felt for suggesting institutional reforms in extension service.
- ☆ The era of online marketing is now influencing innovation in the agrifood sector to compress value chains and innovative business models are delivering higher prices to farmers and higher quality to consumers. This is an innovation space that is evolving rapidly and will bring disruption into existing market mechanisms. Studies on online innovative modes of marketing of farm produce have to be organized.
- ☆ Research institutions need to work with rural entrepreneurs to accelerate the delivery of science-based solutions that can attract youth back to agriculture and treat agriculture as a business. To harness the benefits of scale in in value chains, collaborative and collective action through FPOs is needed.

- ☆ There is need for a Community of Practice (CoP) for Innovation in Agriculture. A Community of practice is a group of people who "share a concern or a passion for something they do and learn how to do it better as they interact regularly". A Community of practice is a support network for people who share a common role. It will create opportunities for learning, building capability, sharing knowledge and reducing duplication of work.

REFERENCES

DAC and FW (2017) Report of the Committee on Doubling Farmers' Income: Vol XI- Empowering Farmers through Extension and Knowledge Dissemination, Ministry of Agriculture and Farmers' Welfare, Government of India.

Devaux, A., Torero, M, Donovan, J and Horton, D (2018) Agricultural Innovation and inclusive value chain development: A review, Journal of Agribusiness in Developing and Emerging Economies, vol 8, No.1., p. 99-123

Kane Gerald, C *et al.* (2021) The Transformation Myth: Leading your organisation into an uncertain future, The MIT Press

Prasad, R M (2017) Action Research on Revisiting SREP and ATMA for effective and sustainable convergence – Concept note submitted to MANAGE, Hyderabad

Utz, Anuja and Dahlman, Carl (2007) Promoting Inclusive Innovation, World Bank

World Bank (2012) Agricultural Innovation Systems: An Investment Sourcebook, Agriculture and Rural Development, World Bank

Annexure 1

Practical Innovation Glossary

The 70-20-10 rule (Eric Schmidt), is a rule for allocating resources between the core (70 per cent), the adjacent (20 per cent) and the transformational (10 per cent) innovation. It's considered to be a good starting for resource allocation. The 70-20-10 rule is closely related to the Three Horizons of Growth model.

Adjacent innovation refers to leveraging the core business and value proposition of an organization in a new market space.

Agile engineering refers to software development practices and methods that embrace change, engage with customers early, and deliver incremental value, frequently.

Architectural innovation is one where the core components of the product remain the same, but the relationship between these components changes. This type of innovation entails the overall design, system or the way components interact.

Breakthrough innovation is a type of innovation that creates new markets. It usually refers to revolutionary change in firms, markets and industries, which provide substantially higher customer benefits relative to current products in the industry.

Business model innovation refers to the ability to find new revenue streams and maintain competitive advantage either by improving an existing business model or by looking for new ways to provide value

Commercialization is a component of product innovation process that converts ideas and prototypes into viable products.

Core innovation refers to small, incremental improvements to existing products or services that are the main sources of revenue, which almost all companies undertake.

Crossing the Chasm is a book written by Geoffrey A. Moore which explores and extends the diffusion of innovations, and is closely related to the technology adoption life cycle. Moore's main argument is that because of very different expectations of early adopters and early majority, there's a chasm between these two that is relatively difficult to cross.

Design thinking is an outcome-focused methodology used for creating better solutions for customers. Design thinking applies actionable design principles to the innovation strategy and ways of working.

Diffusion of innovations is a process by which an innovation is communicated through certain channels over time among the members of a social system. It attempts to identify what aspects influence the adoption rate of an innovation.

Disruptive innovation refers to a new technology that completely changes the way the industry or market functions, creating a new value proposition in an uncontested market space

Fail-fast, Fail-safe: An approach in product development that encourages experimentation and risk taking by bringing decisions early in the process: an unsuccessful result is smoothly accepted as soon as it happens early enough (fail fast), before significant allocation of resources, investment and prior to any dependencies or expectations made (fail-safe).

Hackathon: In practice, it is frequently used to describe any idea-generation initiative with or without functional software deliverable.

Incremental innovation refers to a series of small, gradually built improvements to existing products, processes or methods to maintain competitive position over time.

Innovation is the introduction of something new (Merriam-Webster dictionary). It often includes the development and implementation of a new idea, design, method, product or service.

Innovation culture is essentially an organizational culture that really values and supports innovation, so that people can actually make innovation happen. In practice, it's the engine that drives the organization to constantly get better, move forward, and innovate

Innovation management refers to handling of all the activities needed to innovate, such as creating ideas, developing, prioritizing and implementing them, as well as putting them into practice.

Innovative Organisation: An organization is innovative when it consistently demonstrates an in-creased ability to generate validated Innovation Opportunities in alignment with its purpose.

Innovator's Dilemma is Clayton Christensen's concept introduced in his book that bears the same name. The dilemma describes the phenomenon of new technologies causing established firms to fail in innovation due to certain incentive traps guiding decision-making.

Lean Startup is a methodology for developing businesses ideas and products. It aims to discover if a proposed business model is viable, and shorten product development cycles by adopting a combination of business-hypothesis-driven experimentation, iteration, and validated learning

Marketing innovation refers to an existing product that is used to a whole new purpose and provides a different value proposition for a new customer segment.

Minimum Viable Product (MVP) is the version of a new product a team uses to collect the maximum amount of validated learning about customers with the least effort. It's a basic version with the smallest number of capabilities that will deliver enough value to potential paying customers for them to be able to give you feedback.

Modular innovation or ***component innovation*** refers to innovations where either one or several components of a product system changes, the overall design staying the same.

Open Innovation means combining and using internal and external ideas together to advance the *development* of new technologies. external ideas are used together with the internal ones. Open innovation stresses the importance of relationships between the firm and its outside partners.

Open Innovation 2.0 means that innovation happens in ecosystems or networks that go far beyond traditional organizational boundaries. It's a more open version of open innovation as it widens the scope.

Process innovation refers to the implementation of a new or significantly improved production or delivery method (including significant changes in techniques, equipment and/or software).

Product innovation refers to the development and implementation of a new or significantly improved product.

Radical innovation happens when a new technology completely disrupts existing business or economy and creates a new business model.

Reverse innovation refers to innovations that are first seen and used in emerging markets and only after that introduced to the developed markets.

Social innovations are new ideas or innovation strategies that are designed to meet different social needs, such as health or education.

Sustaining innovation refers to the type of innovation that exists in the current market and instead of creating new value networks, it rather improves and grows the existing ones.

Technology innovation means generating new ideas based on technology, capability or knowledge to produce a new tech solution into a viable entity

Transformational innovation is rare and powerful. It transforms the way organizations do business and offers completely new value for many generations. Only 10 per cent of innovations are transformational.

Value innovation (created by W. Chan Kim and Renée Mauborgne) is a process in which an organization improves, eliminates or changes some of the elements in a product or service to differentiate from competitors by creating a new uncontested market space while also decreasing its costs.

Annexure 2

Tasks of National, State and Sectoral Innovation Councils

National Innovation Council

Formulating a Roadmap for Innovation for 2010-2020

Creating a framework for:

- ☆ Evolving an Indian model of innovation, with focus on inclusive growth
- ☆ Delineating policy initiatives within the Government, required to spur innovation
- ☆ Developing and championing innovation attitudes and approaches
- ☆ Creating appropriate eco-systems and environment to foster inclusive innovation
- ☆ Exploring new strategies and alternatives for innovations and collaborations
- ☆ Identifying ways and means to scale and sustain innovations
- ☆ Encouraging Central and State Governments to innovate
- ☆ Encouraging universities and R&D institutions to innovate
- ☆ Facilitating innovations by SMEs
- ☆ Encouraging all important sectors of the economy to innovate

- ☆ Encouraging innovation in public service delivery
- ☆ Encouraging multi-disciplinary and globally competitive approaches for innovations

State Innovation Council

The objectives of State Innovation Council are:

- ☆ Support State Government to promote innovation in the State,
- ☆ Encourage young talent and local Universities, MSMES and R&D institutes,
- ☆ Map opportunities for innovation in the state
- ☆ Identify and reward talent in innovation and disseminate success stories
- ☆ Create state innovation portal
- ☆ Help create innovation ecosystem
- ☆ Prepare an Innovation Roadmap 2010-2020 for the state

Sectoral Innovation Councils

The objectives of Sectoral Innovation Councils are:

- ☆ Focus on innovation in a specific sector
- ☆ Map opportunities for innovation in the sector
- ☆ Help create a sectoral innovation ecosystem
- ☆ Provide support to promote innovation in the sector
- ☆ Encourage innovations in the public service delivery
- ☆ Prepare a sectoral roadmap for Innovation 2010-2020

Annexure 3

India Innovation Index

The India Innovation Index measures innovation inputs through 'Enablers' and innovation output as 'Performance.' Human capital Investment, Knowledge workers, Business environment and Safety and Legal environment are the five Enabler pillars. Two Performance pillars are- Knowledge output and Knowledge diffusion

Enablers

Human Capital

- ☆ Enrolment in PhD.
- ☆ Enrolment in engineering and technology
- ☆ Colleges connected through NMEICT
- ☆ Higher education institutions with NAAC grade A and above
- ☆ Pupil teacher ratio - Higher education
- ☆ Schools with ICT labs
- ☆ National Achievement Survey (NAS) Scores (Class X)

Investment

- ☆ Expenditure on higher and technical education
- ☆ Expenditure on research and development (state government)
- ☆ Expenditure on science, technology and environment

- ☆ FDI inflow
- ☆ Venture capital deals

Knowledge Workers

- ☆ Knowledge-intensive employment
- ☆ NGOs involved in knowledge intensive areas
- ☆ Private R&D units
- ☆ State funded R&D units

Business Environment

- ☆ Ease of doing business ranking
- ☆ Online services transactions
- ☆ Internet subscribers
- ☆ Incubator centres
- ☆ Common facility centres
- ☆ Cluster Strength

Safety and Legal Environment

- ☆ Information Technology/Intellectual Property related acts
- ☆ Cyber crime police stations
- ☆ Pendency of court cases

Performance

Knowledge Output

- ☆ Grassroot innovations
- ☆ Patent applications filed
- ☆ Trademark applications filed
- ☆ Presence of start-ups
- ☆ Industrial design applications filed
- ☆ New businesses registered
- ☆ Publications

Knowledge Diffusion

- ☆ ICT exports
- ☆ High and medium hi-tech manufacturing entities
- ☆ Geographical Indications (GI) registered
- ☆ Citations

Annexure 4

Global Innovation Index

The indicators of Global Innovation Index (GII) are:

1. **Institutions**
   - ☆ Political environment
   - ☆ Regulatory environment
   - ☆ Business environment

2. **Human Capital and Research**
   - ☆ Education
   - ☆ Tertiary education
   - ☆ Research and Development

3. **Infrastructure**
   - ☆ ICTs
   - ☆ General Infrastructure
   - ☆ Ecological Sustainability

4. **Market Sophistication**
   - ☆ Credit
   - ☆ Investment
   - ☆ Trade, diversification and market scale

5. Business Sophistication

- ☆ Knowledge workers
- ☆ Innovation linkages
- ☆ Knowledge absorption

6. Knowledge and Technology Outputs

- ☆ Knowledge creation
- ☆ Knowledge impact
- ☆ Knowledge diffusion

7. Creative Outputs

- ☆ Intangible assets
- ☆ Creative goods and services
- ☆ Online creativity

Annexure 5

Scheme for Innovation and Agri-Entrepreneurship Development

The components of the scheme are:

- Seed Stage Funding as grant-in-aid to Startups
- Agripreneurship Orientation
- Idea/Pre-Seed Stage Funding

Seed Stage Funding as Grant-in-Aid to Startups

This seed-stage funding will be available to startups who are incubated at the RKVY-RAFTAAR-Agribusiness Incubators.

Financial Assistance- Under this, financial assistance of a maximum of Rs. 25 lakhs will be provided to potential startups that have a minimum viable product (MVP) based on innovative solutions, processes, products, services, and business models in the agriculture and allied sector.

The applicant entrepreneur will be provided funds as per the genuine requirements and as per appraisal/evaluation of their business plans by the RAFTAAR Selection and Monitoring Committee and the decision of the RAFTAAR Selection and Monitoring Committee in this regard will be final.

Objectives

- To ensure timely seed support to deserving entrepreneurs within R-ABIs.
- To enable the translation of minimum viable product (MVP) to the marketable stage and scale up the product and business
- To provide a platform for faster experimentation and modification in their approaches or minimum viable product (MVP) based on innovative solutions, processes, products, services, business models, *etc.* for scaling up.

Eligibility Criteria

- All entrepreneurs of an R-ABI will be eligible for this funding based on consistent performance evaluated by R-ABI Incubation Committee.
- The startup should be a registered legal entity in India with a minimum of two months of residency at the R-ABI.
- The recipient should be DIIPT Recognized startup. This support is not applicable for Indian Subsidiaries of MNCs/foreign companies.

Pattern of Funding Support and Release of Funds

Under the scheme, it is proposed to support around 500 startups during the scheme period. A maximum of 20 startups per R-ABI will be supported under the scheme. Each selected startup will be provided a maximum limit of Rs. 25 Lakh as a grant-in-aid under the scheme. For ensuring Entrepreneurs/ private participation in the program, the grant will be released in instalments.

- The first instalment of 40 per cent of the sanctioned project for seed support will be released after signing of the agreement between the incubator/RABI and the entrepreneur.
- The second instalment of 40 per cent will be released to the entrepreneur after the recommendation of RIC based on the performance and achievement of milestone and after utilization of 80 per cent of the previously released amount and receipt of provisional UC of 80 per cent of already released funds
- The third instalment of 20 per cent will be released to entrepreneurs by the Nodal Division through R-ABIs after the recommendation of RIC based on their performance and achievement of the milestone as well as after utilization of 80 per cent of the previously released amount and receipt of provisional UC of 80 per cent of already released funds.

After two months, market access will be facilitated by the R-ABI and The R-ABI will also organize a Demo Day for Investor pitch to showcase the matured product of the entrepreneur to investors for further facilitation of funding.

Agripreneurship Orientation

The Agripreneurship Orientation program is introduced to provide an opportunity for students to work on their innovative ideas by imparting two months of training cum hands-on internship with other startups at their respective R-ABI. After successful completion of the two months orientation program, the R-ABI will facilitate the incorporation of entities for all viable business ideas. After this program, these startups will be eligible for Idea/Pre-Seed Stage Funding.

Objectives

- ☆ To nurture Agri-entrepreneurs by providing training cum internship with other startups to provide the practical, technical, and business insights
- ☆ To create and nurture a pipeline of Agri-entrepreneurs for incubators
- ☆ To make pursuing entrepreneurship related to innovative ideas an attractive career option among other available career choices

Eligibility Criteria

- ☆ The recipient should propose one innovative idea based on technology, service, business platforms, *etc.* for increasing efficiency in agriculture and allied sector:
  - ❒ Agro-processing
  - ❒ Artificial intelligence
  - ❒ Digital agriculture
  - ❒ Farm mechanization
  - ❒ Waste to wealth
  - ❒ Dairy and fisheries
- ☆ The applicants are expected to pursue their entrepreneurial aspirations full time.
- ☆ The startup should have an initial business plan or proposal for the idea they intend to pursue.

Pattern of Funding Support and Release of Funds

Under the Agripreneurship Orientation, a stipend of Rs 10,000 per month will be granted to these interns for two months during the hands-on training period by the incubators from the Agribusiness Incubator (***R–ABI***) recurring grant.

Idea/Pre-Seed Stage Funding

As part of this scheme, a subsistence grant of a maximum of Rs. 5 lakhs will be provided to interns who have completed the initial two months of the

Agri-entrepreneurship orientation program and are aspiring to convert the innovative idea based on technology, service, business platforms, *etc.* into a prototype/product.

Startups supported for pre-seed stage findings will be eligible for obtaining the next stage of Seed Funding after evaluation.

Objectives

- ☆ To develop an innovative idea based on technology, service, business platforms into prototype/product, services/platforms and launching of a business based on that
- ☆ To provide an opportunity to budding agri-entrepreneurs/students for venturing into the agribusiness arena as well as for minimizing risks involved in businesses based on innovation
- ☆ To attract and retain a large number of enthusiastic and skilled youth in agriculture
- ☆ To create a pipeline of innovative startups for incubators

Eligibility Criteria

- ☆ After successful completion of the initial two months of the agri-entrepreneurship orientation program, interns will qualify for getting Idea or Pre-Seed Stage Funding.
- ☆ The recipient should be registered for the incubation program at the Agribusiness Incubator (***R–ABI***) for the entire duration of Idea/Pre-Seed Stage Funding.
- ☆ The applicant needs to register his/her company before the release of the grant or the startup should be a DPIIT recognized startups

Pattern of Funding Support and Release of Funds

The selected startup will be provided with a maximum limit of Rs 5 lakh as grant in aid under the scheme for working on the idea to convert it into a marketable product. The amount of fund will be released in two instalments as mentioned below:

The first instalment of 60 per cent of the amount of the sanctioned project for support under Idea Stage Funding will be released after signing of the agreement between the incubator/R-ABI and the entrepreneur

The final instalment of 40 per cent will be released to the Entrepreneur through R-ABIs after the recommendation of RIC and RC based on the performance and milestone achieved as well as after utilization of 80 per cent of the previously released amount.

Annexure 6

Intellectual Property (IP) Protection Regulations for Agriculture

1. Geographical Indication (GI)

GI refers to a type of intellectual property protection, which identifies goods with a geographical region. It is a sign or tag used on products having a specific geographical origin and possessing qualities or a reputation due to that origin. GI sign identifies a product as originating in a given place and the qualities, characteristics or reputation of that product are attributed to the place of origin. A GI right enables the right to use the indication to prevent its use by a third party whose product does not conform to the applicable standards. The GI tag is an essential component for creating and maintaining the originality of the product of certain essentials and characteristics. GI plays an important role in promoting the conservation of biodiversity among the rural populations.

The laws relating to the preservation of G.I.s in India are the 'Geographical Indications (Registration and Protection) Act, 1999' (G.I. Act), and the 'Geographical Indications (Registration and Protection of Goods) Rules, 2002 G.I. Rules). India enacted its G.I. law under TRIPS. Under the G.I. Act, since 2003, the Ministry of Commerce and Industry of GOI has established a Geographical

Indication Registry in Chennai, with the jurisdiction of Pan-India, where rights holders can register their G.I.

The registration of GI is a legal process which can be done by any association of persons, producers, organizations or authority established by or under the law who represent the interest of the producers by applying to the GI registry with the prescribed application fee. The registered applicant is considered as the registered proprietor of the product, who is responsible by law to maintain the standards of the production of products. The proprietor organisation also gets authorized user registration along with the GI registration. GI registry will constitute a committee to verify the standards maintained by the producers of the region. For getting GI registration, the indications should fall within the purview of Section 2 (1) of G I Act, 1999. When this happens, it must also meet the provisions of Section 9, which prohibits the registration of a geographical indication.

Proprietorship of the geographic region is treated as a community right. The right to produce the product remains with the producers of the geographical territory. In India, Darjeeling tea was the first product to obtain a GI tag in 2004. Mango, citrus and banana have bagged the highest number of GI tags. Elite clones of fruit crops such as grape, pineapple, litchi, strawberry, guava, kokum, fig and custard apple have also obtained GI tags. Chilli followed by brinjal and onion have got the maximum number of GI tags among the vegetable crops.

The GI used to describe an agricultural product or foodstuff should cover:

- ✰ Originating in a specific region, place or country;
- ✰ Possessing a specific quality reputation or other characteristics attributable to that geographical origin and that product and/or processing of which is done in the defined geo- graphical area; and
- ✰ Existence of any established/traditionally valued direct link between the quality and characteristics of the product and its specific geographical origin.

Rajiv Gandhi National Institute of Intellectual Property Management (RGNIIPM) located at Nagpur under the Ministry of Commerce and Industry organises training programmes for the public on IPR. The farmers can get training on GI from this institute. Unethical marketing practices may undermine the purpose of GI tagging in the absence of an effective post-GI mechanism. This is a major concern, which needs to be addressed to achieve the intended purpose of the GI Act. The GI tags need to be supported by a strong enforcement mechanism both in domestic and export markets.

2. Trademark

Trademark is legally regarded as a type of IP. Trademarks are used in agriculture for marketing of produce or for services. The fundamental aim of a trademark is to differentiate goods and services from one reputed company and to avoid the disappointment of the customers. Trademark registrations are done under the Trade Marks Act, 1999. A mark may consist of a recognizable sign, design or expression which identifies products and services of a particular company and distinguishes it from those of others. The Trade Marks registry is at Mumbai and its branches are located in Kolkata, Chennai, Ahmedabad and New Delhi.

After accession to the Madrid Protocol, a treaty under the Madrid System for international registration of trademarks, the Trade Marks Registry also functions as an office of origin in respect of applications made by Indian entrepreneurs for international registration of their trademarks and as an office of the designated Contracting party in respect of international registrations in which India has been designated for protection of the relevant trademarks.

Section 2(1) (e) of the Indian Trade Marks Act, 1999 defines a trade mark as "a mark capable of distinguishing the goods or services in connection with which it is used in the course of trade certified by the proprietor of the mark in respect of origin, material, mode of manufacture of goods or performance of services, quality, accuracy or other characteristics from goods or services not so certified".

Under the Indian Trade Marks laws, the different types of trade marks that can be registered are a) Product trademarks, b) Service standards, c) Certified trademarks and d) Collective trademarks. The Trade Marks Registry has classified goods and services under 45 classes. Class 31 includes grains, agricultural and horticultural products, livestock products, fruits and vegetables, *etc.*

3. Patent

A patent is an exclusive monopoly right granted by the Government for an innovation to an inventor for his/her disclosed invention for a limited period of time. A request for patent can be made if the innovation meets three requirements- novelty, non- obviousness and marketability/utility.

Compared to developed countries like USA, the number of patent filings in agricultural sector is much less in India. Most of the patent applications from the Indian agricultural sector were filed in agrochemical segments such as biocides, pesticides, herbicides, pest repellents, *etc.*

In 1997, the US Patent office granted a patent to US firm Rice Tec Co (Patent No. 5663484) for a variety called Texmatirice.Texmatiricehas16claims similar to Indian basmati rice. As for GI, basmati rice is an Indian rice variety which could not be patented by other country.

The US permits the broadest variety of agricultural patents. It is has issued regular patents for entire plant lines, such as specific lines of herbicide resistant rice. Such varieties are probably unpatentable in most countries, including India.

Annexure 7

Investment Phases of a Startup

PRE-SEED STAGE (Early stage, Pre-product)

This stage is considered to be the starting point where the founder(s) tries to convert the idea into a business opportunity. This stage requires a small funding for the research where the viability of an idea is assessed. In this stage the possible forces of funding are mostly self-financed (savings), family and friends, angel investors, accelerators and lately a new breed of funding method known as crowdfunding.

SEED STAGE (Early stage, Immature product)

This stage focuses on orienting the company in the broader marketplace and developing a deeper understanding of what the customer wants and how to refine the product according to their taste. The focus of this stage is achieving a viable product market. Another important key aspect of this stage is gaining traction in the target market.

EARLY STAGE DEALS

1. SERIES A (Early stage, Maturing product)

The goal of the Series A stage funding is to position the startup for future growth and get all the cards settled in a row and prepare for expansion in the next stages. By this stage, the idea should be gaining traction and should be creating a good amount of support among the technical people. This is also the stage where business needs to be optimized. During the Series A stage, the startups are working with venture capitalists and angel investors to further refine and improve the original concept

2. *SERIES B (Early stage, Established product)*

By the time startup enters this stage, the company and the products should be fairly well established. The main focus should be on expanding both internally (growing the team) and externally (growing in target audience). Series B investment might allow a startup to make expansive hires, expand into different market segments or experiment with different revenue streams.

LATE STAGE DEALS

SERIES C, D, E.... (Late stage, Fully matured product)

During Series C+ investment, the owners, as well as the investors, are cautious about funding. The more the investment rounds, the more release of the business' equity takes place. Series C rounds are done to fuel large scale expansion, like moving into new market.

BRIDGE LOANS and MEZZANINE FINANCING

(Later stage, Maturing to Mature product)

Bridge financing is when investors invest in a startup business with a short term loan in order to help it reach the next round of funding, on the basis that they will receive their money back. Mezzanine financing is generally offered to companies that have a track record in their industry, an established reputation and product, a history of profitability and a viable expansion plan for the business, such as through expansions, acquisitions or an Initial Public Offering (IPO).

IPO (Late stage, Fully matured product)

An initial public offering of success is a crucial moment for a startup. It is a sign of success. It also comes with its share of costs. This is when the startup becomes a publicly traded company. A successful IPO spells out success for a company. It generates interest and can be a signal to the top talent in the industry that this company has made it.

Annexure 8

Innovations in Agriculture Related to Climate Action Showcased by Sustainable Innovation Forum (SIF)

1. GreenWave

Bren Smith, owner of Thimble Island Ocean Farm and founder of non-profit GreenWave, is a pioneer in the development of regenerative ocean farming. His techniques use 3D ocean farming to create blue carbon, carbon that is captured by the world's ocean and coastal ecosystems.

The GreenWave model is a sustainable, polyculture commercial farming system that regenerates ocean ecosystems by creating carbon and nitrogen sinks, helping to reduce the impacts of ocean acidification. Smith's model consists of vertical farms of scallops, mussels, oysters, clams and seaweed, used for animal feed, fertiliser and to help reduce bioplastics.

2. Symbrosia

The global livestock sector accounts for approximately 14.5 per cent of all greenhouse gas emissions, of which 65 per cent is emitted by cattle. This is where

Symbrosia comes in. Symbrosia is focusing on reducing the methane impact of cows by introducing seaweed into their diets.

The company is scaling the innovative solution to reduce livestock methane emissions using aquaculture techniques. By using sustainable farming techniques like aquaculture, it avoids the degradation of its ecosystems.

3. Aero Farms

Founded in 2004 and based in New Jersey, Aero Farms grows over 800 different crops using an indoor vertical farming system. Unlike hydroponic farming, a technique that replaces soil with nutrient-rich water. Aero Farms use aeroponic technology. This process involves using a cloth made from recycled plastic to mist the roots with water, oxygen and nutrients. Combined with efficient, LED lights designed to have a specific intensity and spectrum, the stable indoor conditions disrupt the cycle of indoor pests, eliminating the need for pesticides, fungicides and herbicides.

4. Trapview

Trapview is an automated pest monitoring and forecasting system that provides farmers with near real-time data about the conditions of their crops. By forecasting potential pest outbreaks, the AI technology serves as a digital employee to help protect the survival rates of crops. Farmers who use Trapview experience lower production costs, higher yields, better risk management and operate under a much more sustainable ecosystem.

Founded in Slovenia in 2007, Trapview manages over 4,000 fields and claims to possess the largest database of algorithms and pest data in the world.

5. Kray Technologies

They have designed the world's first fully digital and unmanned drone crop sprayer. Capable of autonomously spraying up to 1200 acres per day, the agricultural innovation hosts a variety of benefits for the average farmer.

6. Joyn Bio

Founded in September 2017, Joyn Bio was established as a joint venture by Bayer, a pharmaceutical and life science company, and Ginkgo Bioworks – an analytics company that produces their own microorganisms to engineer plant microbes in order to solve the various sustainability issues within the agriculture sector.

Aiming to engineer microbes that provide plants with biological nitrogen fertiliser, Joyn Bio subsequently reduces the amount of industrial nitrogen fertiliser; cutting greenhouse gas emissions by 3 per cent, reducing water pollution and providing late-season nitrogen to crops.

7. BioFiltro

BioFiltro has come up with an unconventional wastewater treatment solution. Founded in Chile in 2009 and beginning operations in USA in 2013, BioFiltro exploits the digestive power of worms and microbes to remove around 99 per cent of bacteria and other contaminants in wastewater. The treatment process allows for water to be reused for agricultural irrigation.

Thanks to their patented Biodynamic Aerobic System, BioFiltro states that they reduce greenhouse gas emissions by 91 per cent - providing a highly sustainable nature-based solution to wastewater treatment within the agriculture sector.

8. Vence Corp

The innovative company designs dynamic boundaries that predict the grazing behaviour of livestock. The Vence app also enables customers to be equipped with a large database about the animals' well-being and general health through lightweight sensors attached to their necks.

According to Vence, the virtual fencing technology helps provide up to 30 per cent cost savings, detect health anomalies and reduce animal management costs.

Index

J

K

L

M

N

O

T

U

V

Y

Z